Soils for Fine Wines

Soils for Fine Wines

ROBERT E. WHITE

OXFORD
UNIVERSITY PRESS

2003

OXFORD
UNIVERSITY PRESS

Oxford   New York
Auckland   Bangkok   Buenos Aires   Cape Town   Chennai
Dar es Salaam   Delhi   Hong Kong   Istanbul   Karachi   Kolkata
Kuala Lumpur   Madrid   Melbourne   Mexico City   Mumbai   Nairobi
São Paulo   Shanghai   Taipei   Tokyo   Toronto

Copyright © 2003 by Oxford University Press, Inc.

Published by Oxford University Press, Inc.
198 Madison Avenue, New York, New York, 10016

www.oup.com

Oxford is a registered trademark of Oxford University Press

Library of Congress Cataloging-in-Publication Data
White, R. E. (Robert Edwin), 1937–
    Soils for fine wines / Robert E. White.
        p.   cm.
    Includes bibliographical references (p.   ).
    ISBN-13: 978-0-19-514102-3

    1. Soils. 2. Viticulture. 3. Wine and wine making. I. Title.
    S591   .W493   2003
    634.8—dc21   2002003662

9 8 7 6 5 4

Printed in the United States of America
on acid-free paper

Preface

Wine making is a very old industry, especially in Europe and West Asia. Dean and Berwick (2001) confirm the common perception that wine in Europe has always been dominated by the soil, specifically through the French concept of *terroir* (Johnson 1994). Books such as Pomerol's (1989) *The Wines and Winelands of France*, described as a "geological journey," and Wilson's (1998) *Terroir: The Role of Geology, Climate and Culture in the Making of French Wines* attest to the central importance of the soil and underlying geology in influencing wine "typicity" and quality (by the European Union [EU] definition). Nevertheless, in New World countries such as Australia, the United States, Chile, South Africa, and New Zealand viticulturalists and wine scientists generally appear to have placed little emphasis on the soil. Books such as Winkler et al.'s (1974) *General Viticulture*, and Iland and Gago's (1997) *Australian Wine from the Vine to the Glass*, both written from a New World perspective, pay scant attention to soil and its complex interaction with wine grapes. However, much emphasis has been placed on climate, as evidenced by Gladstones's (1992) *Viticulture and Environment*, and to viticultural practices, as in Smart and Robinson's (1991) *Sunlight into Wine: A Handbook for Winegrape Canopy Management*. The prevailing view of many in Australia is encapsulated by Smart (2001, p. 48), who concluded that climate has a more important influence on wines than soil because, for example, "there are Terra Rossa soils in the Riverland, yet will anyone suggest a similarity in wine style from there to the Coonawarra?"

Nevertheless, May (1994, p. 34) wrote that "the great interest and unfulfilled need of Australian viticulturalists in knowing more about their soils became evident during the early discussions on the need to publish an Australian textbook on viticulture" (which emerged as *Viticulture: Volume 1 Resources* and *Volume 2 Practices*, edited by Coombe and Dry [1988a,b]). May (1994, p. 34) argued that a "new effort was needed to ensure that viticulturalists have a clear understanding of their soils, of the specific problems they present, and of the best methods to maintain soil productivity," but he found that the treatment of soil in *Viticulture. Volumes 1 and 2* fell short of expectations.

The present book is intended to address the need to provide a clear understanding of soils and their properties, and how those properties influence a soil's response to management for production, specifically of wine grapes. That the soil is an essential component of *terroir* is acknowledged, and quantitative evidence adduced, where available, of how soil, local climate, cultural practices, and variety interact to determine grape quality and the distinctiveness of a wine's character. But more importantly, the book aims to provide a sound understanding of the best practices for managing soils of diverse properties, for the main purpose of optimizing the value of the wines, and also for maintaining satisfactory soil health and physical condition in the longer term. Further, the off-site impacts of intensive viticulture (involving fertilizers, pesticides, herbicides, soil disturbance, and supplementary water for irrigation) may cause environmental problems, if not properly managed. This is particularly true in the main wine-producing areas of southern and southeastern Australia, where maintaining the quantity and quality of water is a serious issue. In the more densely populated areas of the Old and New Worlds, vintners are increasingly being required to demonstrate that wine production does not adversely affect the environment.

This book is not intended primarily as an academic text—rather it is intended for use by viticulturists, wine makers, and consultants to the wine industry in all its facets. It should also appeal to those who appreciate fine wines and want to be better informed about the wide range of soil variation that occurs naturally, and how this variation may influence wine character. Nevertheless, this book should also be a valuable resource for lecturers and students in courses on practical viticulture, viticultural science, and oenology at universities and colleges.

Writing this book would not have been possible without the support and encouragement of Professor Gary Sposito of the University of California (UC), Berkeley, in whose laboratory I spent a most enjoyable time in 1999. I recall many stimulating conversations with him on topics ranging from poetry to wine to complex surface chemistry. I have been fortunate in being escorted to vineyards, and introduced to many enjoyable wines, by a number of people. For this, I am most grateful to Dr. Cornelius van Leeuwen in Bordeaux, Dr. Gabrielle Callot in Montpellier, Professor Francis Andreux in Dijon, Dr. Cliff Ohmart in Lodi, California, Mr. John Livingston of the Napa Valley, Dr. Emilio Ruz-Jerez in Chile, Dr. John Baham of Oregon State University, and Dr. Mike Trought of Villa Maria in New Zealand. Others generous with their advice and time were Professor Larry Williams, University of California (Davis), Mr. Paul Verdegaal and Mr. Ed Weber of the UC Cooperative Extension Service, and Dr. Daniel Roberts of Kendall Jackson, Sonoma. I am most grateful to a number of Australian colleagues and friends who have read one or more chapters and given valuable comments: Dr. Robert Bramley, CSIRO Land and Water; Dr. Peter May, Burnley College; Professor Alex McBratney, University of Sydney; Dr. Tony Proffitt, Southcorp Wines; and Dr. Mike Treeby, CSIRO Plant Industry. However, the final responsibility for the content and conclusions lies with me.

Wine production is truly a vertically integrated industry, in which consumers can relate their appreciation of the product to its specific source. Gathering the information for this book has been a most enjoyable experience, not only for me but also for my family, and especially my wife Annette, who found it some compensation for the many hours I have spent at my computer. I am grateful for their support and also to Oxford University Press for the opportunity to produce this book.

Robert E. White
Melbourne, Australia
January 2002

Contents

Soils for Fine Wines

1   *Soil and the Environment*

1.1   *The Soil and Terroir*

English has no exact translation for the French word *terroir*. But *terroir* is one of the few words to evoke passion in any discussion about soils. One reason may be that wine is one product of the land where the consumer can ascribe a direct link between subtle variations in the character of the product and the soil on which it was grown. Wine writers and commentators now use the term *terroir* routinely, as they might such words as rendezvous, liaison, and café, which are completely at home in the English language.

French vignerons and scientists have been more passionate than most in promoting the concept of *terroir* (although some such as Pinchon (1996) believe that the word *terroir* has been abused for marketing, sentimental, and political purposes). Their views range from the metaphysical—that "alone, in the plant kingdom, does the vine make known to us the true taste of the earth" (quoted by Hancock 1999, p. 43)—to the factual: "*terroir viticole* is a complex notion which integrates several factors . . . of the natural environment (soil, climate, topography), biological (variety, rootstock), and human (of wine, wine-making, and history)" (translated from van Leeuwen 1996, p. 1). Others recognize *terroir* as a dynamic concept of site characterization that comprises permanent factors (e.g., geology, soil, environment) and temporary factors (variety, cultural methods, wine-making techniques). Iacano et al. (2000) point out that if the temporary factors vary too much, the expression of the permanent factors in the wine (the essence of *terroir*) can be masked. The difference between wines from particular vineyards cannot be detected above the "background noise" (Martin 2000). A basic aim of good vineyard management is not to disguise, but to amplify, the natural *terroir* of a site.

Terroir therefore denotes more than simply the relationship between soil and wine. Most scientists admit they cannot express quantitatively the relationship between a particular *terroir* and the characteristics of wine produced from that *terroir*. Nevertheless, the concept of *terroir* underpins the geographical demarcation

3

of French viticultural areas: the Appellation d'Origine Contrôllée (AOC) system, which is based on many years' experience of the character and quality of individual wines from specific areas. There are many examples in France of distinctive wines being produced consistently in particular localities, of which some of the better known are the Grand Cru and Premier Cru of the Côte d'Or in Burgundy, the First and Second Growths of the Médoc in the Bordeaux region, and the vintage champagnes of Champagne.

In spite of the restrictive aspects of the appellation concept (section 9.2.1), this approach has been adopted with various modifications in other European countries. Increasingly, this concept underpins the recognition of American Viticultural Areas (AVA) in the premium wine areas of the Napa and Sonoma Valleys in California. Until recently it had not featured strongly in the viticulture of other countries, such as Australia, where from the earliest times Busby (1825, p. 7) wrote that "with the exception, perhaps, of a very stiff clay, or a rich alluvial loam, there are good vineyards on almost every description of soils." However, where there is a perceived marketing advantage in associating a wine with soil in a specific region, the *terroir* concept is being exploited. One example is the Terra Rossa soil of the Coonawarra region in South Australia, and another the Red Cambrian soil of the Heathcote region in Victoria.

1.1.1   *The Importance of Soil Management in Vineyards*

The role of *terroir* in determining wine character has been comprehensively explored for French vineyards by Pomerol (1989) and Wilson (1998) (for a discussion of terms relating to wine, see box 1.1). Their approach has been to infer the influence of soil on wine through the surrogates of geology and landscape, rather than through properties of the soil itself. This impression is reinforced by the descriptive and cartographic approach to defining different *terroir* in many of the articles of the international colloquium on "Les Terroirs Viticoles" in 1996. For the present, the subtleties of soil factors in determining the distinctive character of wines remain primarily the concern of wine makers supplying the premium end of the market. But good soil management in growing grapes for wine, or "wine growing," is important to all producers for the following reasons:

- The grape vine (*Vitis vinifera*) is a long-lived perennial. Whether grafted on a rootstock or grown on its own roots, each plant is expected to grow for many years. Over the 40- to 50-year life of a commercial planting, with as many as 10,000 plants per ha (as in the Médoc of France), the vines will root deeply and explore virtually all the soil available. This means that the relationship between soil and vine must be as harmonious as possible.
- Most importantly, the soil supplies water and nutrients that influence the vigor of the vine, the balance between vegetative growth and fruit, the yield of grape berries, and berry quality.
- A number of pests and diseases of grapevines, such as nematodes, phylloxera, and the spores of the downy mildew fungus, live in the soil and can have a profound effect on the health and longevity of the plants.
- Maintaining the soil in good condition physically (soil structure), chemically (adequate nutrients and no toxicities), and biologically (organic matter turnover and biodiversity) is important for sustained yield and vine longevity.

| Box 1.1 | *Terms to Describe Wine* |

Quality   An official wine designation in the European Union (EU), which recognizes quality wine as the higher of two general categories of wine. The first category is "quality wine," which must be produced in a specified region defined according to climate and soil type, planting material, cultivation methods, and yield; and the second is "table wine." The quality of wine in each of these categories can vary widely (Robinson 1999). In the New World, quality is not officially defined, but is commonly considered to depend on fruit properties and the wine makers skill (section 9.2.3).

Aroma   A tasting term for a pleasant smell. In wine tasting, *aroma* generally refers to the smells associated with a young wine, as distinct from the more complex aromatic sensations resulting from the aging of a wine. Noble et al. (1987) developed the "aroma wheel" as a standardized approach to the sensory evaluation of wines and to define the terminology objectively.

Bouquet   A French word, adopted in English, to describe a bunch of flowers. Whereas "aroma" relates more to the variety of grape, "bouquet" refers more to the complexity of smells associated with a mature wine (Robinson 1999).

Nose   A versatile word used to mean smell, aroma, or bouquet; a very important component of tasting a wine (Robinson 1999).

Personality   A combination of aroma and bouquet; a complex biological alchemy reflecting the *terroir*, especially the soil and geological attributes (Pomerol 1989).

Character   A synonym for "personality."

- Vignerons must demonstrate that their activities do not adversely affect the environment. Adverse impacts can arise from the use of pest- and disease-control chemicals, fertilizers, and water, and the disposal of waste materials from the winery, through to cultivation that predisposes to soil erosion.

These are some of reasons why grape growers, wine makers, consultants, advisers, and wine writers should have a basic knowledge of the soil on which vines grow. Whether we accept the French view that *terroir*, incorporating the soil, is an all-important determinant of wine character and quality, or the more pragmatic view of the New World that good wines can be produced from most soils, *the soil cannot be ignored as an integral component of wine growing*. This book provides a basic knowledge of soils for viticultural use for those who make wine and enjoy wine, and who have a keen interest in the environment.

1.2   *Basic Concepts of Soil*

1.2.1   *The Soil as an Interface*

The soil is at the interface between the *atmosphere* and the *lithosphere* (the mantle of rocks making up Earth's crust). It also has an interface with bodies of fresh and salt water (collectively called the *hydrosphere*). The soil sustains the growth of many plants and animals, and so forms part of the *biosphere*.

As a result of being situated at these interfaces, the nature and properties of a soil are influenced by the following key factors: weathering of rock materials—

either consolidated rocks such as granite or basalt weathering in place or unconsolidated materials that have been transported; water as rain or snow from the
atmosphere, or water from underground sources (groundwater); and plants and
animals of all sizes that grow or live on and in the soil.

These factors interact in a complex way to form soil. Many soils are very old
(hundreds of thousands of years), so it is difficult to trace all the processes that
have created the soil as seen today. Others, such as those developed on rock materials exposed after the last Ice Age of the Pleistocene epoch (approximately 11,000
years before present or B.P.), are relatively young. In this case we can deduce with
more confidence the processes involved in their formation. Box 1.2 gives a summary of the geological time scale.

A simple example of soil formation under a deciduous forest in the cool humid areas of Europe, Asia, and North America, on calcareous deposits exposed after the last Ice Age, is shown in figure 1.1. The initial state of formation is little

---

Box 1.2	*The Geological Time Scale*

As the science of geology developed, the history of Earth's rocks was
subdivided into a time scale consisting of eras, periods, and epochs. Most
information is available for rocks of the Paleozoic Era, and younger eras, going
back some 560 million years B.P. Periods within the eras are usually associated
with prominent sequences of sedimentary rocks that were deposited in the area
now known as Europe. But examples of these rocks are found elsewhere, so the
European time divisions have gradually been accepted world-wide. A simplified
version of the geological time scale from the Cambrian period to the present is
shown in table B1.2.1.

Table B1.2.1 *The Geological Time Scale*

Era	Period	Epoch	Million Years B.P.
Cenozoic	Quaternary	Recent	0.011
		Pleistocene	2
	Tertiary	Pliocene	5
		Miocene	23
		Oligocene	36
		Eocene	53
		Paleocene	65
Mesozoic	Cretaceous		145
	Jurassic		205
	Triassic		250
Paleozoic	Permian		290
	Carboniferous		360
	Devonian		405
	Silurian		436
	Ordovician		510
	Cambrian		560
Precambrian			560–4600

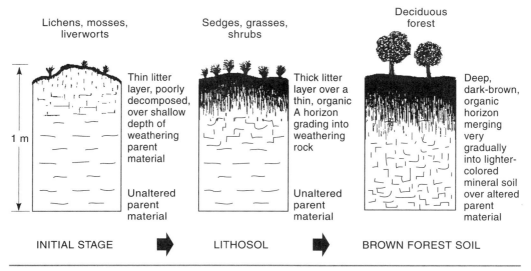

Figure 1.1   Stages in soil profile formation on calcareous parent material in a humid temperate climate (White 1997). Reproduced with permission of Blackwell Science Ltd.

more than a thin layer of weathered material stabilized by primitive plants, such as lichens, mosses, and liverworts. The lichen is able to fix N_2 gas from the air (section 4.2.2.1) and to extract other essential elements for growth from the weathering rock surface. Within a century or so, as the organomineral material accumulates, more advanced species of sedge and grass, which are adapted to the harsh habitat, become established. The developing soil is sometimes called a *Lithosol*. Colonizing microorganisms and animals feed on the dead plant remains and gradually increase in number and variety. Leaf litter deposited on the surface is mixed into the soil by burrowing animals and insects, where its decomposition is hastened by microorganisms. Thus, the remnants of each generation of plants and animals provide the diet of carbon compounds, or substrate, for their successors. The eventual appearance of shrubs and trees, with their deeper roots, pushes the zone of rock weathering farther below the soil surface. After a few hundred years more, a *Brown Forest Soil* emerges.

The names *Lithosol* and *Brown Forest Soil* are examples of soil names that have a broad connotation. Because of the complexity of soil formation, many different soil types occur in the landscape, which inevitably leads to many soil names. Soil variability and the naming of soils are discussed in box 1.3.

1.2.2   *The Soil Profile*

The diagrams of figure 1.1 introduce the concept of a *soil profile*. This is the vertical development of the soil from the surface to the weathering rock below (fig. 1.2). The weathering rock is called *parent material*. Within the soil profile, horizontal zones of different color, stoniness, hardness, texture, and other visible or tangible properties are often visible. These are called *soil horizons*. The upper horizon, from which materials are generally washed downward, is described as *eluvial*;

Box 1.3   *Soil Classification*

Soil scientists have long sought to create order out of apparent chaos. The traditional approach to classification in biology is to allocate "like individuals" to the same class, whereas "unlike individuals" are assigned to other classes. Names and a summary of the key distinguishing properties of each class identify the classes. This is called *classification*.

With plants and animals, the genetic inheritance of traits in individuals is strong and forms the basis of the Linnaean system of classification. The genetic influence in soil formation is weaker and is often obscured by environmental factors and human interference. So soil scientists have devised other ways of classifying soils that take into account, to varying degrees, genesis of the soil, the influence of environmental factors, and actual soil properties. Many systems have been developed, usually for the soils of one country. Other systems purport to be internationally applicable, such as Soil Taxonomy (Soil Survey Staff 1996), promoted by the United States and the FAO-Unesco Classification (FAO-Unesco 1988), promoted by the Europeans. In Australia, the new Australian Soil Classification (Isbell 1996) supersedes the previously used Great Soil Group Classification (Stace et al. 1968). However, many of the highly prized soils for viticulture continue to be called by their Great Soil Group names, or, as in France, by class names modified from the FAO-Unesco Classification. The more common class names, and their cross-correlations, are given in appendix 1. These names are used where appropriate in this book.

All these systems are general-purpose classifications. They have limitations when detailed information is needed about the use of a particular soil for the specific purpose of viticulture. Therefore, it is important always to describe a soil in situ and to measure the key physical and chemical properties of that soil. These data should be stored in a spatially referenced database, such as a Geographical Information System (GIS) (section 8.2.1).

---

lower horizons in which these materials accumulate are called *illuvial*. The notation of A and B is used for the eluvial and illuvial horizons, and C horizon is used for the parent material, as shown in figure 1.2. Organic litter on the surface, not incorporated in the soil, is designated the L layer. Details of soil horizon nomenclature are given in specialist books, such as the Soil Survey Manual (Soil Survey Division Staff 1993) and the Australian Soil and Land Survey Field Handbook (McDonald et al. 1990).

Soil formation or *pedogenesis* is normally more complicated than the simple sequence in figure 1.1 suggests. Many soils have undergone successive phases of development as a result of changes in climate and other environmental factors over time. Such soils are *polygenetic* in origin. In other cases, two or more *layers* of different parent material, created at different times in the geological history of a site, may be found in one soil profile. Soil formation can also be markedly influenced by human activity. Around old settlements, people frequently deposited domestic and animal waste to increase soil fertility. Such soils are usually rich in organic matter and are called *plaggen* soils. Plaggen soils are common in European vineyards.

Figure 1.2   Soil profile showing A, B, and C
horizons. Note the color change
among the sandy loam A horizon,
the clay-enriched B horizon with
Fe_2O_3 accumulation, and the
pale-colored C horizon of
weathering siltstone. Photograph
by the author. See color insert.

1.3   *Factors Affecting Soil Formation*

Jenny (1941) formalized the idea that soil formation at a particular site depended
on the interaction among a soil's inheritance (the parent material), the environment (climate), and organisms (plants and animals), as well as the duration of this
interaction (time). He described these variables as the *soil-forming factors*. The
range of possible climates, parent materials, and values of the other soil-forming
factors is huge, so the range of soil types observed across the landscape is also very
large.

Most soil classifications rely to some extent on a scientific understanding of
how a set of values for the soil-forming factors can produce a particular soil type.
Although this understanding is far from complete, some useful generalizations can
be made about the range of soils produced as any one of these factors varies, on
the assumption that the other factors are constant. We briefly examine each factor in turn.

1.3.1   *Parent Material*

1.3.1.1   *Rock Types*

Consolidated rocks are of igneous, sedimentary, or metamorphic origin. *Igneous
rocks* are formed when molten magma solidifies in or on the earth's crust and are
the ultimate source of all other rocks. Material is released explosively from volca-

Figure 1.3   Granite boulders exposed by weathering in northeast Victoria. Photograph by the author. See color insert.

noes (volcanic ash and pyroclastic flows, collectively called *tephra*), or it flows over the land as streams of lava. Volcanic magma that does not reach the surface accumulates in magma chambers where it solidifies as a pluton or batholith. Subsequent weathering and erosion of the overlying weathered material exposes the harder, more resistant rock as boulders, which remain as dominant features in the landscape (fig. 1.3).

Igneous rocks are broadly subdivided into two categories based on their mineral composition: (1) *acidic* rocks (e.g., granite, rhyolite), which are relatively rich in quartz and the light-colored calcium, potassium, and sodium feldspar minerals, and (2) *basic* rocks (e.g., basalt, gabbro), which are low in quartz, but high in the dark-colored ferromagnesian minerals (hornblende, olivine, pyroxene) (section 2.2.3.2).

Quartz is a very resistant form of crystalline silica (SiO_2). The higher the ratio of Si to (Ca + Mg + K) in a rock, the more likely it is that some of the Si is not bound in silicate minerals such as feldspars, but remains as residual quartz when the rock weathers. Thus, soils formed on granite have more quartz grains than those formed on basalt, and they are usually well drained and deeply weathered (fig. 1.4). Soils formed on basalt are usually high in silicate clay particles and less well drained than soils formed on granite, unless they are very old and hence highly weathered.

Sedimentary rocks are composed of weathered rock products deposited by wind and water. Cycles of geologic uplift, weathering, erosion, and subsequent deposition in rivers, lakes, and seas have produced thick sequences of sediments. Under the weight of overlying sediments (hundreds of meters thick), the deposits gradually consolidate and harden to form rocks. Faulting, folding, and tilting move-

Figure 1.4   A deep, coarse-textured, well-drained soil in the Calquenas region, south-central Chile. Photograph by the author. See color insert.

ments, followed by erosion, can cause the sequence of rock types in the geological column to be revealed by their surface outcrops in a more or less horizontal plane. As the size of the rock fragments decreases, sedimentary rocks change from *conglomerates* to *sandstones* to *siltstones* to *mudstones*. During periods of active mountain building, erosion rates are very high and much sedimentary material of mixed particle size is deposited. The French term for this is *roches molassiques* or molasse, which is well represented in the Bordeaux, Armagnac, and Cognac regions of southwest France.

Other sedimentary rocks such as limestone and chalk are formed by precipitation from solution or from detrital material (e.g., the calcareous skeletons of marine algae). Limestones vary in composition from pure calcium carbonate ($CaCO_3$) to mixtures of calcium and magnesium carbonate (dolomite), or carbonates with much sand, silt, and clay. As such rocks weather, the $CaCO_3$ dissolves, and the soil that is formed consists of resistant mineral impurities mixed with residues of organic matter. Soils on limestone or chalk are therefore usually shallow and well structured because of their high Ca content (chapter 3). Figure 1.5 shows a *rendzina*, a typical soil formed on hard limestone in McLaren Vale, South Australia, where it is used for growing grapes.

Metamorphic rocks are originally igneous or sedimentary rocks that have been subjected to heat and pressure. There are two broad types of metamorphism: contact or thermal metamorphism, and regional metamorphism. Contact metamorphism occurs when magma intrudes into sedimentary rocks. At the higher temperatures around the zone of intrusion, minerals in the original rock are

Figure 1.5   A Rendzina soil profile in McLaren Vale, South Australia. Photograph by the author. See color insert.

transformed into more resistant forms, resulting in a harder and denser rock. An example is *quartzite*, which is formed by the recrystallization of quartz grains in sandstone. Regional metamorphism occurs when rocks in Earth's crust are subjected to increased temperature and pressure over millions of years. New minerals and rock types are formed, such as when mudstone is progressively metamorphosed (depending on the intensity of the process) to *shale, slate, schist,* or *phyllite.* Figure 1.6 shows weathered schist near St. Chinian in Languedoc-Roussillon, southern France, which supports vines up to 100 years old. Similar schist occurs in Haut-Beaujolais. Limestone, including dolomite, is metamorphosed to *marble* (as in the Dolomitic Alps of Northern Italy).

1.3.1.2   *Transported Parent Materials*

Water is the dominant agent in *weathering*, not only because it initiates mineral dissolution and hydrolysis, but also because it sustains plant life on rock surfaces. Lichens play a special part in weathering, because they produce "chelating agents" (section 1.3.3.1), which trap elements from the decomposing rock in organometal complexes. Carbon dioxide from the decay of plants and other organisms enhances the solvent action of rainwater by forming weak carbonic acid. Plant roots also contribute to the physical disintegration of rock initiated by temperature changes and ice wedges.

Most of the earth's surface has undergone several cycles of submergence, uplift, erosion, and denudation over many millions of years. During the mobile and depositional phases, much mixing of materials from different rock formations took place. Transported parent materials are therefore usually very heterogeneous, and the pattern of soil formation on these materials is correspondingly complex. The agents of transport are *water, ice, wind,* and *gravity.*

Figure 1.6   Old vines growing in weathered schist in Languedoc-Roussillon, southern France. Photograph by the author. See color insert.

Water carrying suspended rock fragments abrades surfaces. During transport, rock material is sorted according to size and density and abraded, so that waterborne or *fluviatile* deposits are characteristically smooth and rounded sand particles, gravel, pebbles, or boulders, depending on their size. Transport by water produces alluvial and terrace deposits. Figure 1.7 shows a young soil used for vines in the Borden Ranch district near Lodi in California's Central Valley. This soil is forming on gravelly alluvial material, probably representing two main phases of active deposition, and is very freely drained. The larger fragments are either rolled or bounced along the stream bed by a process called *saltation*. The smallest particles that remain in suspension are called *colloidal*. These require a long time to settle and hence form sedimentary deposits only in deep, calm water (siltstones and mudstones).

Ice was an important agent in the transport of rock materials during the 2 million years of the Pleistocene (box 1.2). During the colder glacial phases, the ice cap advanced from polar and mountainous regions to cover a large part of the land surface, especially in the Northern Hemisphere. The moving ice ground down rock surfaces, and the "rock flour" was incorporated in the ice. Rock debris also collected on the surface of valley glaciers by *colluviation* (see below). During the warmer interglacial phases, the ice melted and glaciers retreated, leaving extensive deposits of heterogeneous glacial drift or *till*. Streams flowing out of the glaciers produced glaciofluvial deposits of sands and gravels in which the pebbles are less smooth and rounded than in fluviatile deposits. Figure 1.8 shows an example of a glaciofluvial deposit on top of older weathering granite near Calquenas in south-central Chile. These soils are increasingly being used to grow wine grapes, although maintaining the water supply to the vines can be difficult because of the soil's free drainage.

Figure 1.7   River gravel deposits used for viticulture in the Lodi District, Central Valley, California.
                    Photograph by the author. See color insert.

Figure 1.8   Glaciofluvial deposit on older
                    weathering granite in the
                    Calquenas region, south-central
                    Chile. Photograph by the author.
                    See color insert.

Wind moves rock fragments by rolling, saltation, and aerial suspension. Material swept from dry periglacial regions during the Pleistocene has formed deposits called *loess* in the central United States, central Europe, northern China, and Argentina. These deposits are many meters thick in places and form the parent material of highly productive soils. Similar windblown deposits called *parna* are common in parts of the southern Murray-Darling Basin in Australia. Where the parna cover is relatively thin, the underlying buried soils, typically Red Brown Earths or Red Earths, may be revealed. Soils on shallow and deep parna are used for irrigated viticulture along rivers in the Griffith, Sunraysia, and Riverland districts of the Murray-Darling Basin.

Gravity produces colluvial deposits as fragments of weathered rock slide down steep slopes in mountainous regions. Less obvious, but widespread under periglacial conditions, are *solifluction* deposits at the foot of slopes. These deposits are formed when frozen soil thaws from the surface downward and the saturated soil mass slips over the frozen ground beneath. Many vineyards are planted on solifluction deposits in valleys and on escarpment slopes in France, Italy, and Germany, as, for example, the vines growing on colluvial deposits derived from limestone in the Côte d'Or. Slips and larger earth flows can occur on steep slopes on unstable parent materials when forest vegetation is cleared and replaced by vineyards.

1.3.2   Climate

Climate has a major effect on soil formation and is also an important determinant of *terroir* in wine growing. Various climatic classifications have been developed to indicate regions most suitable for particular grape varieties. These have been reviewed by Gladstones (1992). Smart and Robinson (1991) recognize three levels of climate:

1. *Macroclimate* or regional climate: usually applicable over tens of kilometers, depending on topography and the distance from moderating influences, such as the sea
2. *Mesoclimate* or site climate: more local than a macroclimate, determined by altitude, slope, and aspect, such as, for example, the south-facing steep slopes in the Mosel Valley in Germany. In California, the macroclimate of the Lodi District in the Central Valley is different from the macroclimate of the Napa Valley, but within each region, there are several mesoclimates based on distance from San Francisco Bay, closeness to the mountains, and elevation.
3. *Microclimate* or canopy climate: the climate within and immediately around the vine canopy

The key components of climate for soil formation are *water* and *temperature*.

1.3.2.1   Water

The effectiveness of water in soil formation depends on the form and intensity of the precipitation, its seasonal variability, the evaporation rate (from vegetation and soil, see chapter 6), land slope, and permeability of the parent material.

Leaching. In wet or humid climates, there is a net downward movement of water in the soil most of the year. This downward flow or *drainage* usually results

Box 1.4   *The Mobility of Elements in Soil*

The mobility of an element depends on its solubility in water and the effect of pH on that solubility. Relative mobilities, represented by the Polynov series in table B1.4.1, have been established from a comparison of the composition of river waters with that of igneous rocks in the catchments from which they drain. The low mobility of Al and Fe is explained by the formation and strong adsorption of hydroxy-Al and hydroxy-Fe ions at low pH, and the precipitation of insoluble Al and Fe hydroxides at higher pHs. The least mobile element is titanium, which forms the insoluble oxide TiO_2. Therefore, titanium is used as a reference element to estimate the relative gains or losses of other elements in the profile.

Table B1.4.1. *The Relative Mobilities of Rock Constituents*

Constituent elements or compounds	Relative mobility[a]
Al_2O_3	0.02
Fe_2O_3	0.04
SiO_2	0.20
K	1.25
Mg	1.30
Na	2.40
Ca	3.00
SO_4	57

[a]Expressed relative to chloride taken as 100

in greater leaching of soluble materials, which can be removed from the soil entirely. However, in arid climates where evaporation generally exceeds rainfall, there is a net upward movement of water most of the time, which carries salts to the soil surface.

The extent to which salts are retained in the soil profile depends on the mobility of an element (box 1.4) and the rate of drainage through the soil. In arid areas, even the most soluble constituents, mainly sodium chloride (NaCl) and to a lesser extent the chlorides, sulfates, and bicarbonates of Ca and Mg, tend to be retained and give rise to *saline soils* (section 7.2.2). However, in more humid climates, there is a greater loss of salts and SiO_2, and the soils are more highly leached, except in the low-lying parts of the landscape where the drainage waters and salts accumulate (section 1.3.4).

Lessivage. Downward movement of water can mechanically wash, or translocate, clay particles from the A to B horizon. In humid temperate climates, on acidic parent materials, such lessivage is frequently associated with the leaching of organic compounds from the litter layer. These compounds form complexes with iron (Fe) and aluminum (Al) in the A horizon. The net result is that minerals, such as the iron oxides, which color the soil yellow to red, are dissolved from the A horizon and deposited in the B horizon. The lower part of the A horizon (A2) becomes pale or "bleached" below the upper part (A1), which is darkened by the presence of organic matter. Such soils are called "podzolized," after the Russian word *podzol* which means "ash-colored." These soils are common vineyard soils in southeastern Australia, where they are often referred to as "duplex soils" to indicate the contrast in clay content between the A and B horizons. Figure 1.2 is an example of a duplex soil.

Lessivage is unlikely to occur in calcareous soils or in neutral soils that retain a high proportion of exchangeable Ca^{2+} ions, because the clay is not readily translocated. Under acid conditions, if exchangeable Al^{3+} and hydroxy-Al ions are predominant on the clay surfaces, the clay should also remain flocculated (section 4.5.2). However, if the Al and Fe are complexed by soluble organic compounds and removed, the clay particles are more likely to deflocculate and be translocated. Similarly, when Na^+ ions comprise more than 6–15% of the exchangeable cations, depending on the soil, clay deflocculation and translocation are likely to occur. The result is a *sodic soil* (section 7.2.3).

1.3.2.2 Temperature

Temperature varies with latitude and altitude and with the absorption and reflection of solar radiation by the atmosphere. Temperature controls the state of water—ice versus liquid versus vapor—which has a key effect on soil formation through rock weathering, leaching, and the transport of materials. Temperature affects the rate of mineral weathering and synthesis, and the biological processes of growth and decomposition. Reaction rates are roughly doubled for each 10°C rise in temperature (section 3.5.4), but enzyme-catalyzed reactions are sensitive to high temperatures and usually attain a maximum between 30 and 35°C.

Temperature decreases by 0.5°C per 100 m increase in altitude. On land close to large bodies of water, temperature varies less from day to night and with the seasons than in arid continental regions. This temperature moderation influences soil formation and affects viticulture through time of flowering, fruit set and ripening, and the incidence of frost. The interaction between soil temperature and color is also important for vineyards in cool climates (section 3.5.2).

1.3.3 Organisms

Soil and the organisms living on and in it comprise an *ecosystem*. The active participants in the soil ecosystem are plants, animals, microorganisms, and humans.

1.3.3.1 Plants and Animals

Natural vegetation and soil form a feedback loop because the soil influences the type of vegetation present and the vegetation in turn influences soil formation. A good example of this interaction is the different effects of coniferous trees (pines, spruce, and larch) and temperate deciduous species (oak, elm, ash, and beech) on soil formation. The leaves and leaf litter of coniferous trees are richer than deciduous trees in organic compounds called polyphenols that are powerful reducing and complexing agents (box 1.5). Polyphenols that do not chelate and leach with Fe and Al are polymerized and contribute to the thick litter layers that build up under pine forests. This layer is called *mor humus* (fig. 1.9a). Humus describes organic matter that is in varying stages of decomposition as a result of microbial activity. One reason for the accumulation of mor humus is that the litter of pines, spruce, and larch is unpalatable to earthworms. Mor humus is only slowly broken down by small insects and colonized by slow-growing fungi, so it is poorly

Box 1.5   *Complex Formation and Podzolization*

As indicated in the discussion of lessivage, polyphenols can form soluble organometal complexes or *chelates* with the normally immobile Al^{3+} and Fe^{3+} cations. Typically, ferric iron (Fe^{3+}) is reduced to ferrous iron (Fe^{2+}), with the concurrent formation of a soluble ferrous-organic complex. Under anaerobic conditions, decomposition of the organic compound supplies the electrons that drive the reduction process. Because H^+ ions are also involved in the reduction, the formation of the complexes is favored by low pH. Thus, chelation and subsequent translocation of Fe and Al are more obvious on acidic parent materials and under wet surface conditions. The overall process, called *podzolization*, is best developed under conifers (and heath-type species) in wet climates on acidic, sandy soils.

decomposed. Conversely, the litter of elm and ash, and to a lesser extent oak and beech, is more readily ingested by earthworms and becomes mixed with soil as fecal material. The result is a more uniform distribution in the A horizon, to a depth of 20–30 cm, of well-decomposed organic matter that is called *mull humus* (fig. 1.9b).

Earthworms are the most important of the soil-forming fauna in viticultural regions supported both by small insects and larger burrowing animals (rabbits, moles, and voles). Because of their burrowing, earthworms create large pores that facilitate soil drainage, and so are very important in vineyard soils. The relation-

Figure 1.9   Soil with (a) mor humus and (b) mull humus (White 1997). Reproduced with permission of Blackwell Science Ltd.

(a)

(b)

ship between earthworm numbers and cover crops in vineyards is discussed in chapter 7.

Earthworms are also important in soil formation in warmer regions with moist conditions. But generally in warm regions, the activities of termites, ants, and dung-eating beetles are of greater significance, particularly in the semiarid regions of Africa, Australia, and Asia.

1.3.3.2   *Human Influence*

Humans have influenced soil formation through agriculture and settlement for several thousand years. For example, continual cultivation to 20–25 cm transforms the humus-enriched A horizon (Ah) of an undisturbed soil to a featureless "plow layer" (Ap horizon) in which organic and mineral matter is well mixed. The natural soil structure is also greatly modified. Clearing the native vegetation changes the course of soil formation. Cultivation for crops and grazing of grasslands predisposes the soil to erosion. In some cases, the A horizon may be completely removed, leaving the less fertile B horizon exposed, as in the case of many duplex soils in Australia. Soils on limestone and chalk are particularly vulnerable because on such parent materials, the soil profile is generally shallow and the rate of soil accumulation very slow. Figure 1.10 shows vines growing on limestone fragments with no soil cover in Languedoc-Roussillon, France.

Even under intensive cultivation, a soil profile can be enriched through the input of organic residues and sometimes fine mineral materials. For example, the thin soils on Cretaceous Chalk in the Champagne region of France have been augmented for centuries with lignite from local Tertiary clays and silts. The lignite (called *cendres noires* or black ash) is a soft, low-grade coal that contains iron pyrite,

---

Figure 1.10   Vines growing on limestone fragments near St Jean de Minervois, Languedoc-Roussillon. Photograph by the author. See color insert.

FeS_2. When exposed to air, the pyrite oxidizes to produce sulfuric acid that reacts with the chalk, and makes Fe more available to the vines (section 5.6.2). Also, mineral matter in the lignite improves the soil texture. In more recent years, composted organic wastes from the villages of Champagne have been added as mulch to build up the soils.

Artificial drainage also modifies soil formation. Soils with hydromorphic features form wherever water collects in natural depressions or where soils remain wet for long periods as a result of high rainfall and low evaporation rates. Characteristic hydromorphic features are (1) an accumulation of organic residues because of a very slow rate of decomposition in wet or cold conditions or both, which eventually gives rise to *peat*, and (2) the chemical reduction of Fe^{3+} to more soluble Fe^{2+} compounds, which produces a *gleyed* soil (typically blue-grey in color, as in the clay subsoils of the Pomerol and parts of the Médoc, Bordeaux; see section 9.3.2).

On removal of excess water by drainage, Fe^{2+} iron is reoxidized to the more stable Fe^{3+} form, which precipitates as localized, bright orange and red hydrated oxides, to give a characteristically *mottled* profile. Changes in soil due to drainage are discussed further in section 1.3.4.

1.3.4   *Relief*

1.3.4.1   *Catenas*

The term *catena* (Latin for a chain) describes a suite of contiguous soil types extending from hilltop to valley bottom. The parent material of a soil catena may be the same, with the soil profile differences due entirely to variations in drainage. Other catenas may be on slopes carved out of two or more superimposed rock formations. The soils on the upper slopes and escarpment are usually old and denuded, whereas those on the lower slopes are younger, reflecting the rejuvenation of the in situ soil by materials derived from upslope.

Where the parent material is uniform and permeable, the soils at the top of the slope are freely drained with the water table at considerable depth; whereas in the valley bottom the soils are poorly drained, with the water table near or at the soil surface (fig. 1.11a). As drainage conditions deteriorate, the oxidized soil profile at the top, with its uniform, warm orange-red colors, is progressively transformed into an imperfectly drained mottled soil, and finally into the peaty, gleyed waterlogged soil at the bottom (fig. 1.11b).

1.3.4.2   *Slope and Aspect*

Subtle changes in local climate and vegetation are associated with the slope and aspect of valley sides and escarpments. At higher latitudes, slopes facing the sun dry out more quickly than the reverse slopes. This affects the type and density of vegetative cover, which in turn affects runoff. The steepness of the slope also affects drainage and runoff. Soil moves on slopes, so a midslope site is continually receiving material from sites immediately upslope by wash and creep, and continually losing material to sites below. In this case, the form of the slope is important, whether it is smooth or uneven, convex or concave, or broken by old river terraces. Solifluction deposits under escarpments and on slopes are common in Europe, as in the Rhine, Saône, and Rhone Valleys.

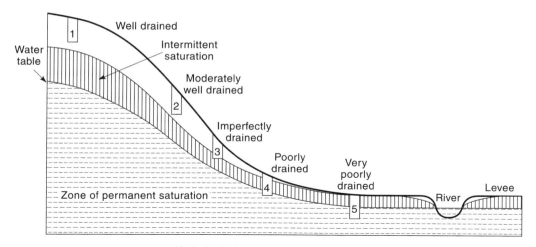

Hydrological sequence of soils from 1 to 5

(a)

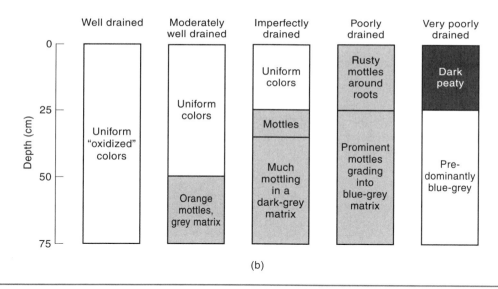

(b)

Figure 1.11　　(a) Section of a slope and valley bottom showing soil drainage classes. (b) Soil profiles, with changing morphology, that correspond to the drainage classes of Figure 1.11a (White 1997). Reproduced with permission of Blackwell Science Ltd.

The gradient of a slope (expressed as a percent) can be measured from a contour map (as the vertical interval divided by the horizontal distance between contours). The slope angle α (alpha) is related to the gradient by

$$\text{Gradient} = \tan \alpha \qquad (1.1)$$

The angle α is measured with an inclinometer.

1.3.5   *Time*

The climate of the earth has changed over geological time. The most recent large changes were associated with the alternating glacial and interglacial phases of the Pleistocene. These phases were accompanied by a rise and fall in sea level, by erosion and deposition, and by corresponding adjustments in Earth's crust, all of which produced radical changes in the distribution of parent material, the vegetation, and the shape of the landscape. On a shorter time scale of only a few thousand years, changes can occur in the biotic factor of soil formation, as shown by the succession of plant species on a weathering rock surface (fig. 1.1). The relief factor also changes through changes in slope form and the distribution of groundwater. During an even shorter period of time, parent material can change as a result of volcanic eruptions.

The Concept of Steady State. When the rate of change of a soil property with time is negligibly small, the soil is said to be in steady state with respect to that property. For example, for a soil forming on a gentle slope, the rate of natural erosion may just counteract the rate of rock weathering so that the soil appears to be in steady state. However, because pedogenic processes do not all operate at the same rate, soil properties may not attain steady state at the same time. Soil formation is not linear with time. In soils of temperate regions, the accumulation of C, N, and organic P follows a "law of diminishing returns," whereas clay accumulation follows a sigmoidal trend (fig. 1.12).

Because colonization of weathering parent material by plants and animals is an integral part of soil formation, steady state should be considered in the context of the soil–plant ecosystem. Studies of such ecosystems on parent material exposed by the last retreat of Pleistocene ice (ca. 11,000 years B.P.) suggest that a stable combination of soil and vegetation (in the absence of human intervention) is achieved in 1,000–10,000 years. The system is stable in the sense that any changes occurring are immeasurably small in the period since scientific observation began (ca. 150 years). The soil in such systems is sometimes referred to as *mature.*

Figure 1.12   The approach to steady state for soil N, organic P, and clay content under temperate conditions (White 1997). Reproduced with permission of Blackwell Science Ltd.

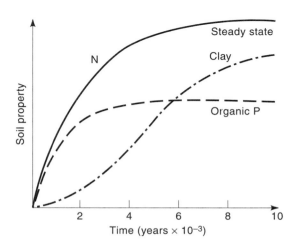

By contrast, in other regions where the land surface has been exposed for much longer (10^5–10^6 years), the attainment of steady state by a soil–plant ecosystem is more problematic. In such old landscapes, the effect of time is confounded by changes in other factors: climate, parent material, and relief. If the rate of mineral weathering is faster than in cool temperate regions, and if leaching is more severe, then weathering occurs to a much greater depth. The deep horizons are less influenced by surface organic matter. Clay minerals break down mainly to iron and aluminum oxides, with the release of SiO_2 and basic cations (chapter 4). Silica not removed in solution remains as resistant quartz grains or hardened layers of silcrete. Under such conditions, climate exerts less influence than parent material and relief on the end product of soil formation.

1.4   *Soil as a Natural Body*

1.4.1   *Current Attitudes about Soil*

During the twentieth century, the predominant attitude regarding soil use was one of exploitation. Exploitation of soil for production has accelerated since the 1940s with the introduction of greater mechanization, expansion of irrigation, and increased use of chemical fertilizers. Although this intensification of agriculture has had the very desirable benefit of enabling world food production to more than keep pace with the increase in population (now greater than 6 billion), there have been costs in the form of worsening soil and water degradation (table 1.1).

Viticulture has the potential to degrade soil and water resources because many vineyards are clean-cultivated and situated on slopes that are sometimes very steep. Soil eroded in runoff can be transported to streams and rivers. But in the Old World, a tradition of nourishing the soil to nurture the vine has developed over centuries, as in Bordeaux and Burgundy (chapter 9). Also, in the New World,

Table 1.1   **Causes, Extent, and Severity of Land Degradation Globally**

Process	Area (Million ha)[a]	Reversibility	Off-site Effects
Water erosion	700	Difficult	Important
Wind erosion	280	Difficult	Important
Loss of nutrients	135	Easy[b]	Possibly important
Loss of organic matter	General	Easy[b]	Negligible to possibly important
Soil acidification	10	Easy[b]	Negligible
Salinization	80	Difficult	Possibly important
Pollution	20	Difficult	Possibly important
Physical damage	60	Difficult	Possibly important

[a]Total area of arable land is ca. 1500 million ha.
[b]Provided the necessary inputs are available and are economically justified
Source: Data from Greenland et al. (1998)

vineyards close to centers of population, as in the Napa and Willamette Valleys, operate under the watchful eyes of environmentally aware communities. In the production of wine, soil and water must be used in such a way that they are not degraded and off-site effects are minimized. To achieve this goal, the vignerons as well as the community must be better educated about soil—its productive potential, its role as a reserve of biodiversity, and as a foundation for stable landscapes.

Thus, a responsible approach to soil use recognizes soil as a *natural body*, clearly distinguished from inert rock by (1) the presence of plant and animal life, (2) a structural organization that reflects the action of pedogenic processes, and (3) a capacity to respond to inputs and changes in the environment, which alter the balance between gains and losses in the profile and determine the formation of a different soil in dynamic equilibrium with new conditions. The third point indicates that soil has no fixed inheritance because it depends on the conditions prevailing during its formation. Nor is it possible to unambiguously define the boundaries of the soil body. The soil air is continuous with the air above the ground. Many organisms can live as well on the soil surface as they can in the soil, and the litter layer usually merges gradually with soil organic matter. Likewise, the boundary between soil and parent material is difficult to demarcate. It is common, therefore, to speak of the soil as *a three-dimensional body that is continuously variable in time and space*. We discuss the scale at which soil variability occurs and its effect on the expression of *terroir* in chapter 9.

1.4.2   *Components of the Soil*

The soil is a porous body comprising about 50% solid matter by volume, with water and gases competing for the remaining pore space. The relative proportions of the four major components—*mineral matter, organic matter, water, and air*—generally lie within the ranges shown in figure 1.13. Soil water contains dissolved organic and inorganic solutes, and is called the soil solution. Although the soil air consists primarily of nitrogen gas (78% by volume), oxygen (21%), and argon (0.9%), it also contains higher concentrations of carbon dioxide (~0.1%) than

Figure 1.13   Proportions of the main soil components by volume (White 1997). Reproduced with permission of Blackwell Science Ltd.

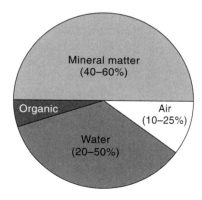

the atmosphere because of respiration by roots and microorganisms (section 3.4). In addition to CO_2, traces of other gases such as methane (CH_4), nitric oxide (NO), and nitrous oxide (N_2O), the by-products of various microbial metabolic processes (chapter 5), are produced.

1.5   ***Summary Points***

This chapter explores the relationship between soil and its environment. The main points are summarized as follows:

- Soil forms at the interface of the atmosphere and consolidated or loose rocks of the earth's crust. Physical and chemical weathering, denudation and redeposition, plus the action of colonizing plants and animals mold a distinctive soil body. The process of soil formation culminates in a remarkably variable differentiation of soil material into a series of *horizons* that make up a *soil profile*. The intimate mixing of mineral and organic materials to form a porous fabric, permeated by water and air, creates a favorable habitat for plant and animal life.

- The five soil-forming factors are *parent material, climate, organisms, relief*, and *time*. Where only one set of factors has operated during soil formation, the resultant soil is said to be *monogenetic. Polygenetic* soils, which are much more common, arise when there is a significant change in one or more of the factors. Studies on the role of these factors, and their interactions, have helped scientists understand why certain types of soil occur where they do. This concept has also been the basis of early attempts at *soil classification*. But modern systems of soil classification rely as much on the measurement of actual soil properties as on the inferred genesis of a soil.

- *Parent material* affects soil formation primarily through the type of rock minerals, their rate of weathering, and the weathering products. These products are redistributed by the action of *water, wind, ice*, and *gravity*. Generally, the parent material of soils formed on transported material (especially alluvium) is more heterogeneous than that of soils formed on weathering rock in situ (residual soils). Many vineyards of the world are located on alluvial deposits (old river terraces), colluvial deposits (rock slides), and solifluction deposits.

- *Climate* affects soil formation through rainfall, evaporation, and temperature. Climate also influences grape-growing directly. Climate interacts with *organisms*, particularly through its effect on the type of vegetation.

- *Organisms* are the vegetation and the animals that live on and in the soil. Human activity can exert a profound effect on the soil, for good or bad, through the clearing of natural vegetation and the use of soil for agricultural and other purposes.

- *Relief* and its effects are obvious in old, weathered landscapes. Slope has an important effect on runoff and soil drainage, which is important for viticulture. Drainage in turn influences the balance between oxidation and reduction of Fe compounds in the soil profile.

- *Time* is important because the value of a soil-forming factor may change with time (e.g., global climate change), and profile differentiation depends on the length of time that a soil has been forming. A soil profile may attain a *steady state* if the rate of natural erosion matches the rate of weathering of the parent material. Soil–plant ecosystems developing since the last Pleistocene ice phase appear to have attained steady state within 10,000 years. On older land surfaces, where the rate of mineral weathering may be faster and leaching more severe, steady state soil–plant ecosystems are less likely to be achieved.

■ Soil is a fragile component of the environment. Its use for food and fiber production and for waste disposal must be managed in a way that minimizes off-site effects and preserves the soil for future generations.

2 *The Makeup of Soil*

2.1 *The Solid Phase*

Minerals and organic matter comprise the solid phase of the soil. The geological origin of the soil minerals, and the input of organic matter from plants and animals, are briefly discussed in section 1.2.1. A basic knowledge of the composition and properties of these materials is fundamental to understanding how a soil influences the growth of grapevines.

2.2 *Soil Mineral Matter*

A striking feature of soil is the size range of the mineral matter, which varies from boulders (>600 mm diameter), to stones and gravel (600 to >2 mm diameter), to particles (<2 mm diameter)—the *fine earth* fraction.

The fine earth fraction is the most important because of the type of minerals present and their large surface areas. The ratio of surface area to volume defines the *specific surface area* of a particle. The smaller the size of an object, the larger is the ratio of its surface area to volume. This can be demonstrated by considering spherical particles of radius 0.1 mm, 0.01 mm, and 0.001 mm (1 micrometer or micron, μm). The specific surface areas of these particles are 30, 300, and 3000 mm^2/mm^3, respectively. In practice, the specific surface area is measured as the surface area per unit mass, which implies a constant particle density (usually taken as 2.65 Mg/m^3). A large specific surface area means that more molecules can be adsorbed on the surface. Representative values for the specific surface areas of sand, silt, and clay-size minerals are given in table 2.1. Note the large range in specific surface area, even for the clay minerals, from as little as 5 m^2/g for kaolinite to 750 m^2/g for Na-montmorillonite.

Because specific surface areas are important, we need to know the size distribution of particles in the fine earth fraction. This is expressed as the soil's *texture*. The types of minerals that make up the individual size fractions are also impor-

Table 2.1   *Specific Surface Areas According to Particle Size and Mineral Type*

Mineral or Size Class	Specific Surface Area (m^2/g)
Coarse sand	0.01
Fine sand	0.1
Silt	1.0
Kaolinite	5–100
Hydrous micas (illite)	100–200
Vermiculites and mixed-layer minerals	300–500
Montmorillonite (Na-saturated)	750[a]
Iron and aluminum oxides	100–300
Allophanes	1000[b]

[a]The maximum value for completely dispersed montmorillonite particles of 1-nm thickness

[b]Includes internal and external surfaces of the mineral

tant because they too influence the reactivity of the surfaces. Both these topics are discussed here.

2.2.1   *Assessing Soil Texture*

All soils show a continuous distribution of particle sizes, called a frequency distribution. This distribution relates the number (or mass) of particles of a given size to their actual size, measured by the diameter of an equivalent sphere. When the mass of particles in each size class is summed sequentially, we obtain a cumulative frequency distribution of particle sizes. Some examples are given in figure 2.1. It is convenient to divide the continuous distribution into several broad class intervals that define the size limits of the *sand, silt,* and *clay* fractions. The class limits vary to some extent between engineers and soil scientists, and from country to country. The major systems in use are those adopted by the U.S. De-

Figure 2.1   Cumulative distributions of particle sizes for a typical clay, sandy silt loam, and sandy soil (White 1997). Reproduced with permission of Blackwell Science Ltd.

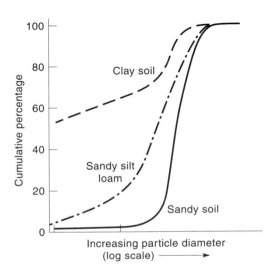

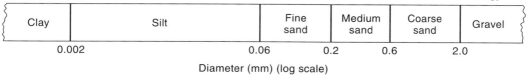

Figure 2.2   Particle size classes most widely adopted internationally (White 1997). Reproduced with permission of Blackwell Science Ltd.

partment of Agriculture, or USDA (the Natural Resource Conservation Service), the British Standards Institution, and the International Society of Soil Science. These are illustrated in figure 2.2.

2.2.1.1   *Particle Size Analysis*

Particle size distribution can be measured accurately in the laboratory or estimated "by feel" in the field. The principles of the laboratory analysis are outlined in box 2.1. The outcome of the laboratory analysis can be summarized in a *textural triangle* of the kind shown in figure 2.3. Depending on the proportions of sand, silt, and clay, the soil is assigned to a textural class. For example, a soil with 40% sand, 40% silt, and 20% clay is classed as a loam in the USDA system. The main differences between the International System, which was adopted in Australia, and that used in the United States lie in the upper size limit for silt and in the number of subdivisions in the sand fraction (see fig. 2.2).

2.2.1.2   *Field Texture*

In the field, a soil surveyor assesses soil texture by moistening a sample with water until it glistens. The sample is then kneaded between fingers and thumb until the aggregates are broken down into individual particles, which must be thoroughly wet. The proportions of sand, silt, and clay are estimated according to the following qualitative criteria:

- *Coarse sand* grains are large enough to grate against each other and can be detected individually by sight and feel.
- *Fine sand* grains are much less obvious, but when they comprise more than about 10% of the sample they can be detected by biting the sample between the teeth.

Box 2.1   *Particle Size Analysis in the Laboratory*

The analysis is carried out on the <2-mm sieve fraction of air-dry soil. This is soil of very low water content that has been dried in the air at 40°C. The soil aggregates must be completely disrupted and the clay particles dispersed. Methods used include treatment with hydrogen peroxide (H_2O_2) to oxidize organic matter that binds particles together, mechanical agitation, ultrasonic vibration, and dispersion of the clay in "Calgon" (sodium hexametaphosphate). Details are given in specialist books such as Klute (1986).

Once a soil suspension has been made, the particles are allowed to settle in a special cylinder, and the density of the suspension is measured at certain times. The coarse and fine sand particles settle out very rapidly, and they are measured by subsequent sieving. The settling velocity v is directly proportional to the particle radius r squared (assuming the particles are spheres), according to the equation

$$v = Ar^2 = \frac{h}{t} \tag{B2.1.1}$$

A is a coefficient that depends on the acceleration due to gravity, the viscosity of water (which changes with temperature), and the densities of the particle and of water. Clay particles with an upper size limit of 2 μm will fall slowly at a particular velocity, from which their depth h in the suspension can be estimated after time t. By measuring the density (mass/volume) of the suspension at that time, the quantity of clay can be calculated. Density is measured by placing a Bouyoucos hydrometer in the suspension or by calculating the loss in weight of a bulb of known volume when immersed in the suspension—the *plummet balance method*.

Constant temperature should be maintained. Two assumptions simplify the calculation of settling velocity using equation B2.1.1:

1. The clay and silt particles are smooth spheres, even though they have irregular flat shapes.
2. The particle density does not vary with mineralogy.

In practice, we use an average particle density of 2.65 g/cm^3, and we speak of the "equivalent spherical diameter" of the particle size being measured. The results of particle size analysis are expressed as the mass of the individual fractions per 100 g of oven-dry soil (<2 mm). Oven-dry soil is soil dried to a constant weight at 105°C. By combining the coarse and fine sand fractions, the soil may be represented by one point on the textural triangle of figure 2.3.

- *Silt* grains cannot be detected by feel, but their presence makes the soil feel smooth and silky, and only slightly sticky.
- *Clay* is characteristically sticky, although dry clays, especially the expanding-lattice types (section 2.2.4.2), require much moistening and kneading before they develop maximum stickiness.
- A high organic matter content reduces stickiness in clays and makes sandy soils feel more silty.
- Finely divided calcium carbonate also gives a silty feeling to a soil.

Depending on the estimated proportions of sand, silt, and clay, the soil is assigned to a textural class according to a textural triangle (fig. 2.3).

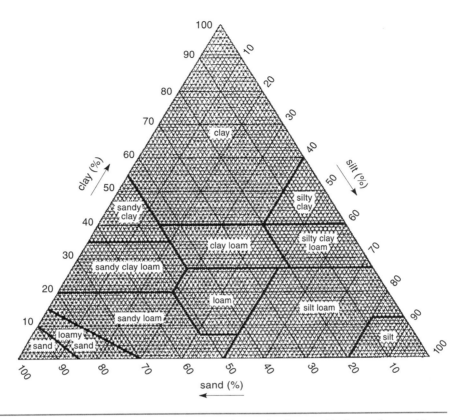

Figure 2.3   A textural triangle based on the USDA particle-size classification (similar to that used in Europe) (Soil Survey Manual 1993). Reproduced with permission of the USDA.

2.2.2   *Soil Texture and Vineyard Soils*

Texture influences soil behavior in several ways, notably through its effect on soil structure, water retention, aeration, drainage, temperature, and the retention of nutrients.

2.2.2.1   *Texture and Soil Water*

Clay particles are very important for aggregation and hence soil structure, which in turn influences water holding capacity, aeration, and drainage. Good aeration and drainage are important for vines. Fine- and medium-textured soils, such as clays, silty clays, and clay loams, are called "heavy textured": they have higher water holding capacities than sandy soils. But unless they are well structured, they are unlikely to drain as quickly as sandy or "light-textured" soils. Gravel and stones make a sandy soil even more free-draining, but also lower the soil's water holding capacity. An example of a very freely drained, gravelly soil used for vine growing in the Lodi District of California was shown in figure 1.7.

| Box 2.2 | *Soil Consistence* |

Consistence is measured by the resistance offered to breaking or deforming a soil when a compressive, shear force is applied. In the field, consistence is assessed on aggregates 20–30 mm in diameter as they are pressed between the thumb and forefinger. Usually the test is done on both air-dry and moist soil (at the "field capacity," as defined in section 3.3.3). The resistance of the aggregate to rupture or deformation is expressed on a scale ranging from loose (no force required, separate particles present) to rigid (cannot be crushed under foot by the slow application of a person's body weight).

A soil that has a high resistance to deformation is said to be strong. High soil strength may be associated with compaction, which is generally undesirable. Detection of compacted layers in vineyard soils is discussed in chapter 7.

2.2.2.2   *Texture and Soil Temperature*

Texture has a pronounced effect on soil temperature. Clays hold more water than sandy soils. Water modifies the heat required to change a soil's temperature because its specific heat capacity is three to four times that of the soil solids, and considerable latent heat is either absorbed or evolved during a change in the physical state of water, for example, from ice to liquid water or vice versa. Thus, the temperature of a wet clay soil responds more slowly than that of a sandy soil to changes in air temperature during spring and autumn (section 3.5.2).

The temperature regime of vineyards can be influenced by the stoniness of the soil surface. Large stones act as a heat sink during the day and slowly reradiate this heat energy during the night, creating a more favorable microclimate in the vine rows. A good example of this effect is in the famous Chateauneuf-du-Pape appellation of the Southern Rhone Valley in France.

2.2.2.3   *Texture and Tilth*

Texture should not be confused with *tilth*, which refers to the surface condition of cultivated soil prepared for seed sowing—how sticky it is when wet and how hard it sets when dry. Tilth therefore depends on the soil's *consistence*, which reflects the influence of both texture and structure, modified by the water content (box 2.2).

2.2.3   *Minerals of the Sand and Silt Fractions*

2.2.3.1   *Basic Crystalline Structures*

Sand and silt particles are mainly resistant residues of *primary* rock minerals, together with small amounts of *secondary* minerals (salts, oxides, and hydroxides) formed by rock weathering. The primary minerals are mainly silicates whose crystal structure is based on a simple molecular unit, the *silicon tetrahedron*, SiO_4^{4-} (fig. 2.4a). Another basic molecular unit found in the aluminosilicate minerals is the *aluminum octahedron*, $Al(OH)_6^{3-}$ (fig. 2.4b). The electric charge properties of these units are discussed in box 2.3. Note that the edges of these molecular units measure only a fraction of a nanometer, which is one-billionth of a meter, so we are examining here the detailed fine structure of a crystal lattice.

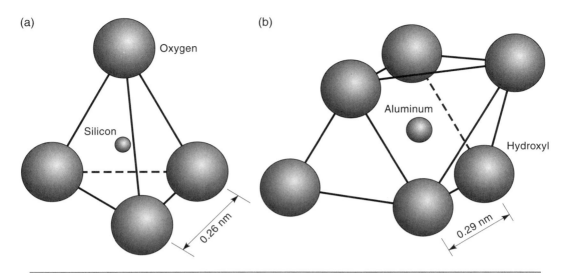

(a)   (b)

Oxygen

Aluminum

Silicon

Hydroxyl

0.26 nm

0.29 nm

Figure 2.4   (a) Diagram of a Si tetrahedron. (b) Diagram of an Al octahedron (interatomic distances not to scale) (White 1997). Reproduced with permission of Blackwell Science Ltd.

2.2.3.2   *Chains and Sheets*

Silicate minerals are formed by linking together the basic silicon and aluminum units (a process of polymerization). This occurs by covalent bonding of O atoms to form chain, sheet, and three-dimensional structures. In the *pyroxene* minerals, for example, each Si tetrahedron is linked to adjacent tetrahedra by sharing two

---

Box 2.3	*Electric Charges and Valency*

   Oxygen in its ionic form O^{2-} has two "free" electrons. Normally, the free electrons are shared between two O atoms to form an uncharged molecule, O_2. When four O^{2-} ions are coordinated to one Si^{4+} ion to form a tetrahedron (fig. 2.4a), the deficit of electrons associated with Si is satisfied by four electrons from the oxygens. But this leaves four negative charges potentially unneutralized. Such a charge imbalance cannot exist in nature, so cations such as Al^{3+}, Fe^{3+}, Fe^{2+}, Ca^{2+}, Mg^{2+}, K^+, and Na^+ are attracted to the SiO_4^{4-} units. The cations become covalently bonded to the O atoms so that the surplus charges on the O atoms in the SiO_4^{4-} unit are neutralized. An electrically neutral silicate crystal called a *feldspar* is formed.

   The number of unneutralized charges (+ or −) associated with the ionic form of an element defines its valency state. Elements of similar size and the same valency frequently substitute for one another in a silicate structure—a process called *isomorphous substitution*. The structure remains electrically neutral. However, when elements of similar size but different valency exchange, there is an imbalance of charge. The most common substitutions are Mg^{2+} or Fe^{2+} for Al^{3+} in octahedral sheets, and Al^{3+} for Si^{4+} in tetrahedral sheets (see section 2.2.3.2). The excess negative charge is neutralized by the incorporation of additional cations into the crystal lattice or by structural arrangements that allow an internal compensation of charge (e.g., in the chlorites).

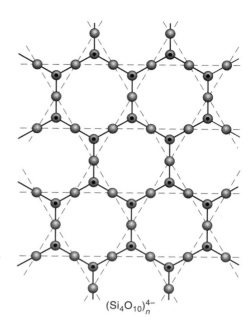

$$(Si_4O_{10})_n^{4-}$$

Figure 2.5   A silica sheet in plan view showing a pattern of hexagonal holes (White 1997). Reproduced with permission of Blackwell Science Ltd.

out of three basal O atoms to form a single extended chain. In the *amphiboles*, two parallel pyroxene chains are linked by the sharing of an O atom in every alternate tetrahedron. Cations such as Mg^{2+}, Ca^{2+}, Al^{3+}, and Fe^{2+} are bonded to the O atoms to neutralize the surplus negative charge. These minerals are collectively called *ferromagnesian* minerals and are found in basic rocks such as basalt, gabbro, and serpentinite.

A one-dimensional chain extended in two dimensions forms a *silica sheet*, consisting of tetrahedra linked by the sharing of all the basal O atoms (fig. 2.5). The apical O atoms (superimposed on the Si atoms, as seen in fig. 2.5) form bonds with metal cations in adjacent sheets by, for example, displacing OH groups from their positions around a trivalent Al^{3+} ion, as in the unit $Al(OH)_6^{3-}$. When Al octahedral units (shown in fig. 2.4b) polymerize by sharing basal OH groups, they form an *alumina* sheet. Such sheets are the basis of the mineral *gibbsite* $[Al_2OH_6]_n$, in which only two-thirds of the available cation positions are occupied by Al atoms. If Mg^{2+} is present instead of Al^{3+}, all the available octahedral positions are filled and the mineral formed is *brucite* $[Mg_3(OH)_6]_n$. When two silica sheets sandwich one alumina sheet, the result is a covalently bonded 2:1 layer-lattice mineral, characteristic of the *micas* (*muscovite* and *biotite*). When mica-type layers sandwich a brucite layer, the resultant mineral is *chlorite*.

These are examples of *phyllosilicate* minerals, common in rocks and in the clay fraction of soils (section 2.2.4). Biotite and chlorite are typical minerals of basic rocks. Some characteristics of phyllosilicate (or layer-lattice) structures are given in box 2.4.

2.2.3.3   *Three-dimensional Structures*

The most important silicate minerals of this type are silica and the feldspars. Silica minerals consist of polymerized Si tetrahedra of general composition $(SiO_2)_n$. Silica occurs as the residual primary mineral *quartz*, which is very inert, and as a

Box 2.4   *Model Layer-lattice Structures*

Phyllosilicates are composed of two-dimensional silica and alumina sheets stacked in regular arrays in the *c* direction. Figure B2.4.1 shows the generalized structure of a phyllosilicate. The following terminology is used:

- A *plane* comprises covalently bonded atoms (such as O and OH) one atom thick.
- A *sheet* is a combination of planes of atoms (such as a silica sheet).
- A *layer* is a combination of sheets (such as two silica sheets combined with one alumina sheet, as in mica).
- A *crystal* is made up of one or more layers.
- Planes of atoms are repeated at regular distances in multilayer crystals, which gives rise to a *d* spacing, or *basal spacing*, characteristic of a particular mineral.
- Between the layers is the *interlayer* space, which may be occupied by water and solute molecules.
- Phyllosilicates generally have a large planar surface and small edge surfaces.

Figure B2.4.1 A model layer-lattice crystal structure (White 1997). Reproduced with permission of Blackwell Science Ltd.

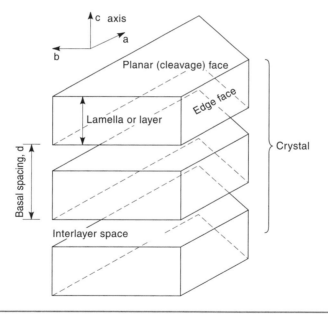

secondary mineral formed by weathering. Secondary silica initially exists as amorphous opal that dehydrates over time to form microcrystalline quartz, such as *flint* or *chert*. Black flints are found in Cretaceous Chalk beds underlying vineyard soils of the middle Loire Valley in France. As the $CaCO_3$ dissolves during weathering, the very durable flints remain embedded in the clay impurities of the Chalk to form a shallow clay-with-flints soil. The insolubility of quartz and flint means that these minerals are abundant in the sand and silt fractions of many soils.

The molecular structure of feldspars is introduced in box 2.3. Their structure consists of polymerized Si tetrahedra in which some Si^{4+} is replaced by Al^{3+}. The cations balancing the excess negative charge are commonly K^+, Na^+, and Ca^{2+} and less commonly, Ba^{2+} and Sr^{2+}. They are chemically more reactive and more easily weathered than silica minerals, and so rarely comprise more than 10% of the sand fraction of mature soils.

2.2.4   *Minerals of the Clay Fraction*

These minerals are subdivided into the *crystalline clay minerals,* predominantly phyllosilicates, and *accessory minerals*, which are salts, oxides, carbonates, and resistant primary minerals reduced to a very small size by weathering. Because of their large specific surface areas and surface charges (table 2.1), these minerals provide very important sites for reactions with nutrients and water in soil (see chapters 4, 5, and 6). The clay minerals are conveniently classified on the basis of their Si:Al mole ratios.

2.2.4.1   *Minerals with a Si:Al Mole Ratio ≤ 1*

The most common mineral of this group is *kaolinite*, which is found in many highly weathered soils. Kaolinite has a 1:1 layer-lattice structure formed by the sharing of O atoms between a silica sheet and an alumina sheet. The basal spacing of the crystals is fixed at 0.72 nm as a result of "hydrogen bonding" between the H and O atoms of adjacent layers (figure 2.6). The layers stack fairly regularly in the *c* direction to form large crystals 0.05–2 μm thick. Because of the hydrogen bonding between layers, water and solute molecules do not penetrate the interlayer spaces, and the specific surface area is low (table 2.1). Kaolinite clays show minimal shrinkage or swelling with a change in water content. The mineral *halloysite* has the same structure as kaolinite, except for the presence of two sheets of water molecules between crystal layers, which causes the layers to curve and form a tubular crystal. Halloysite is found in weathered volcanic ash soils.

Figure 2.6   Structure of a kaolinite crystal with hydrogen bonding between layers (White 1997). Reproduced with permission of Blackwell Science Ltd.

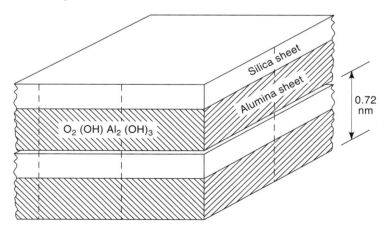

Other minerals in this group are *imogolite* and *allophane*, most commonly found in relatively young soils (1000–2000 years) formed on volcanic ash and pumice. The Si:Al mole ratio ranges from 0.5 in imogolite to 1 in some of the allophanes. These minerals form hollow tubes (imogolite) or spheres (allophane) between 1 and 5 nm in diameter. Because they are hollow, these minerals have very high specific surface areas, encompassing both internal and external crystal surfaces (table 2.1).

Isomorphous substitution of Al^{3+} for Si^{4+} in these minerals is variable, ranging from a few percent in kaolinite to 50% in some allophanes. This substitution gives rise to a negative lattice charge, measured in moles of charge (box 2.5). The lattice charge is neutralized when cations in solution are attracted to the clay surface. A charge within the lattice is considered *permanent* because it depends only on the crystal structure. In addition, a *variable* charge develops at the crystal edge faces due to the association or dissociation of protons at exposed O and OH groups. The combination of charge effects—one due to substitutions within the crystal lattice and the other due to unsatisfied valencies at crystal edge faces—determines the *cation exchange capacity* (*CEC*) of a clay mineral (section 4.6.1).

2.2.4.2 *Minerals with a Si:Al Mole Ratio of 2*

Clay minerals in this group all have 2:1 layer-lattice structures. They differ mainly in the extent and location of isomorphous substitution in the crystal lattice. The main isomorphous substitutions are Al^{3+} for Si^{4+} in the tetrahedral sheets and Mg^{2+} and/or Fe^{2+} for Al^{3+} in the octahedral sheets. The predominant location of the substitution determines the type of cation that is attracted to the mineral to become an interlayer cation in the crystal. The physicochemical reasons for this are discussed in box 2.6. The end result of these ionic substitutions is three main

---

Box 2.5 *Moles of Charge and Equivalents*

The molar mass of an element is defined as the number of grams weight per mole (abbreviated to mol) of the element. The standard is the stable C-12 isotope of carbon. On this scale, H has a molar mass of 1 g, K a mass of 39 g, and Ca a mass of 40 g. The recommended unit of charged mass for cations, anions, and charged surfaces is the *mole of charge*, which is equal to the molar mass divided by the ionic charge. Thus, for the elements H, K, and Ca, the mass of a mole of charge is

H^+ $1/1 = 1$ g
K^+ $39/1 = 39$ g
Ca^{2+} $40/2 = 20$ g

For clay minerals and soils, the most appropriate unit of measurement is the *centimol (cmol) of charge* (+) *or* (-) *per kg*. In the older soil science literature, the charge on ions and soil minerals was expressed in terms of an *equivalent weight*, which is the atomic mass (g) divided by the valency (and identical to a mole of charge). The *CEC* of a mineral was expressed in milliequivalents (meq) per 100 g, which is numerically equal to cmols charge/kg.

Box 2.6   *Cation Adsorption on Clay Surfaces*

Negative charge located at a tetrahedral site (in a silica sheet) is much closer to an adsorbed cation than is a charge at an octohedral site (in an alumina sheet). This means its polarizing effect on the cation is greater and can overcome the force of attraction between the cation and water molecules in its hydration shell. If the cation loses its hydration shell, it can approach the surface very closely to become strongly adsorbed. The ease with which a cation can be stripped of its water molecules and become strongly bound depends on its hydration energy. For the common exchangeable cations, the energy of hydration per mole of cation falls in the order

$$Al^{3+} > Mg^{2+} > Ca^{2+} > Na^+ > K^+ \cong NH_4^+$$

K^+, which has a relatively low hydration energy, is incorporated as an unhydrated cation in the interlayers of mica-type clays, where the lattice charge is predominantly in the tetrahedral sheet. The interlayer K^+ pulls the adjacent mica layers close together. However, when the lattice charge is buried more deeply in the crystal layers, as in montmorillonite, there is less tendency for adsorbed cations to dehydrate and draw adjacent mineral layers close together. Adsorbed cations such as Ca^{2+} remain partially hydrated and form complexes with the surface, as in the case of Ca-saturated montmorillonite. Adsorbed cations such as Na^+ and Li^+ remain fully hydrated and do not complex with the surface.

2:1 mineral types, *illite*, *vermiculite*, and *smectite*, whose chemical formulas are summarized in table 2.2.

A true soil mica such as *illite* has isomorphous substitution predominantly in the tetrahedral sheet. The permanent negative charge is neutralized by unhydrated K^+ ions that fit snugly in the interlayer spaces between adjacent crystal surfaces. The interlayer bonding is therefore very strong, and the basal spacing is between 0.96 and 1.01 nm. But as illite weathers, the interlayer K^+ is gradually replaced by cations such as Ca^{2+} and Mg^{2+}. These have higher hydration energies and remain partially hydrated. The basal spacing expands to 1.4–1.5 nm, which is equivalent to a double layer of water molecules between the mineral layers. Consequently, the interlayer bonding becomes weaker, and the stacking of the layers to form crystals is much less regular. Minerals of this type are sometimes called *hydrous micas*.

Table 2.2   *Examples of the Chemical Composition and Moles of Charge for 2:1 Clay Minerals*

Mineral	Typical Unit Cell Formula[a]	Moles of Charge/Unit Cell
Illite (hydrous mica)	$[(Si_{7.1}Al_{0.9})^{IV}(Al_{3.3}Mg_{0.7})^{VI}O_{20}(OH)_4]^{-y}\,yK^+$	$y = -1.6$
Vermiculite	$[(Si_{7.0}Al_{1.0})^{IV}(Al_3Mg_{0.5}Fe_{0.5})^{VI}O_{20}(OH)_4]^{-y}\,0.5yCa^{2+}$	$y = -1.5$ to -2.0
Smectite (montmorillonite)	$[Si_8^{IV}(Al_{3.2}Mg_{0.6}Fe_{0.2})^{VI}O_{20}(OH)_4]^{-y}\,0.5yCa^{2+}$	$y = -0.6$ to -0.8

[a]The superscripts IV and VI indicate, respectively, whether the atoms are in the tetrahedral sheet (4 O around each Si) or octahedral sheet (6 O or 6 OH around each Al); the Fe is shown as Fe^{2+}, but can also be Fe^{3+}.
Source: Data from Sposito (1989) and White (1997)

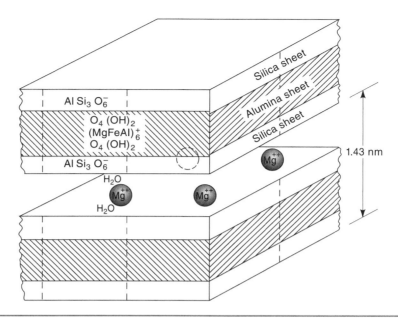

Figure 2.7   Structure of vermiculite crystals. The basal spacing is fixed in the presence of partially hydrated Mg^{2+} ions in the interlayer spaces (White 1997). Reproduced with permission of Blackwell Science Ltd.

Vermiculites have both di- and trivalent cations occupying all the available sites in the alumina sheet, producing a net positive charge that partly neutralizes the negative charge developed through substitution of Al^{3+} for Si^{4+} in the silica sheet. The polarizing effect of the tetrahedral charge is lessened, so that the cations Ca^{2+} and Mg^{2+} in the interlayers are only partially dehydrated. The basal spacing of a Mg-vermiculite, for example, is typically 1.43 nm, as shown in figure 2.7. This collapses to 1 nm on heating to drive off the water, so the vermiculites show limited reversible swelling. In an acid environment, hydrated Al^{3+} ions replace Ca^{2+} and Mg^{2+} in the interlayers and "islands" of $Al(OH)_3$ form as the Al^{3+} ions hydrolyze and polymerize. This gives rise to an aluminous *chlorite* type of clay mineral.

Smectite minerals have a varied composition, with isomorphous substitution occurring in both the silica and alumina sheets. The most common smectite in the soil, *montmorillonite*, has substitution in the alumina sheet only, usually Fe^{2+} or Mg^{2+} for Al^{3+} (table 2.2). The interlayer cations are also partially or fully hydrated (see box 2.6) and freely exchangeable with other cations in solution. Depending on the type of exchangeable cations, the basal spacing may vary from 1.5 to 4 nm, as illustrated in figure 2.8. Ca-montmorillonite, a form common in the soil, has a basal spacing of 1.9 nm at full hydration, when there are three layers of water molecules between the crystal layers. The stacking of the layers is very irregular and the average crystal size is much smaller than in the micas and kaolinites. The interlayer surfaces provide a large internal area that adds to the external planar area, so the specific surface area of smectite clay is very large (table 2.1).

Because of the weak interlayer bonding and the free movement of water and cations in and out of this region, the smectites are called *expanding-lattice clays*.

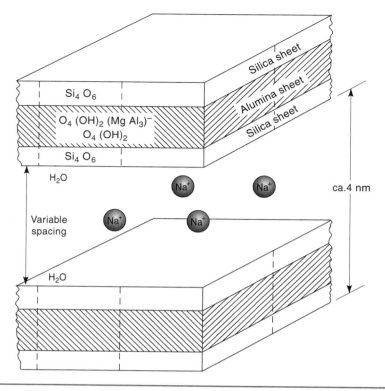

Figure 2.8   Structure of Na-montmorillinite crystals. The basal spacing is variable because of the Na$^+$ ions in the interlayer spaces (White 1997). Reproduced with permission of Blackwell Science Ltd.

Soils with such clays show pronounced shrinkage and cracking when dry and swelling on rewetting (fig. 2.9)

2.2.4.3   *Accessory Minerals*

Accessory minerals are predominantly *free oxides* and *hydroxides*, including various forms of silica, iron oxides, aluminum oxides, manganese, and titanium oxides. Of these, the oxides of Fe, Al, and Mn are the more important.

Iron oxides accumulate in soils that are highly weathered, especially those derived from more basic rocks. They occur as discrete particles or as thin coatings on clay minerals. Iron oxides have strong colors ranging from yellow to reddish-brown to black, even when disseminated through the soil profile. Commonly, when solutions rich in Fe^{2+} seep from waterlogged soils or rock fissures, a voluminous rusty-red precipitate of *ferrihydrite* forms. If conditions remain moist and cool, ferrihydrite slowly transforms to the yellow mineral *goethite* (α-FeOOH). This is the most common and stable iron oxide in the soil. Under hot and dry conditions, ferrihydrite transforms to *hematite* (α-Fe$_2$O$_3$). Hematite is a bright red oxide characteristic of soils that are highly weathered and subjected to periods of intense drying.

Figure 2.9   A vineyard soil with montmorillonite clay that shows surface cracking when dry. Photograph by the author.

Aluminum oxides have a greyish-white color that is easily masked in soils, except where large concentrations occur, as in bauxite ore bodies. In acid soils, precipitates of $Al(OH)_3$ form in the interlayers of chlorites, vermiculites, and smectites, and generally as surface coatings on clay minerals. This poorly ordered material slowly crystallizes to *gibbsite* (γ-$Al(OH)_3$), the principal aluminum oxide mineral in soil.

Several oxides of manganese are found in the soil, ranging from manganous oxide (MnO) to manganese dioxide (MnO_2), according to the valency state of the Mn. The most stable form of MnO_2 is *pyrolusite*, but the most common Mn oxides in the soil are in the *birnessite* group, in which Mn^{2+}, Mn^{3+}, and Mn^{4+} ions are bonded to O^{2-} and OH^-. Small black deposits (1–2-mm diameter) of manganese oxides are common in soils that experience alternating aerobic and anaerobic conditions. The surfaces of these minerals have a high affinity for heavy metals, especially cobalt (Co) and lead (Pb). The availability of Co to plants is often controlled by the solubility of Mn oxides (section 5.6.2).

The Fe and Al oxides have the general formula $(Fe, Al)_2O_3.nH_2O$. Because the mole ratio of oxygen to metal is 3:2, the generic term for these oxides is the *sesquioxides*. The oxides have a variable surface charge, which is usually positive at acid to neutral pHs. Thus they form very stable coatings on negatively charged clay minerals. The positive surface charge also means that they adsorb negatively charged organic and inorganic ions from solution, which confers an *anion exchange capacity* (*AEC*) on the soil. The influence of the sesquioxides on soil properties is discussed in chapters 3 and 4.

2.3   *Organic Matter and Soil Organisms*

2.3.1   *Inputs of Plant and Animal Residues*

Plant residues include aboveground litter fall (leaves and branches) and below-ground root material. Animal residues include dead bodies and excreta. Below-ground, root tips produce mucilage that lubricates the root as it grows through the soil. Epidermal cells produce exudates, and as the cells mature they are abraded and sloughed off, releasing their contents into the soil. Thus, belowground C inputs comprise dead root material, "sloughed-off" root cells, and C compounds exuded into the surrounding soil, which is described as the *rhizosphere*. The total quantity of C compounds released is 5–8% of the total plant C, excluding C released as respired CO_2 (Tinker and Nye 2000). The return of C in animal corpses and excreta is obviously highly variable in type and amount. Table 2.3 gives examples of C inputs and C:N ratios of materials from different sources.

2.3.1.1   *Mineralization and Immobilization*

All organic material is a potential substrate for soil organisms, most importantly the microorganisms that derive energy for growth from the oxidation of complex organic molecules. During decomposition, essential elements are converted from organic to simple inorganic forms through a process called *mineralization*. For example, organically combined N, P, and S appear as NH_4^+, $H_2PO_4^-$, and SO_4^{2-} ions, and more than half the C is released as CO_2. The remaining substrate C is synthesized into microbial cell substances that form the *microbial biomass*. A variable proportion of other essential elements such as N, P, and S is also incorporated, making these elements unavailable for plant growth until the organisms die and decay. This process is called *immobilization*. The residues of the organisms, together with the more recalcitrant parts of the original substrate, accumulate in the soil as *soil organic matter* (*SOM*).

Table 2.3   **Annual Plant Litter and Animal Returns and C:N Ratios**

Source	Organic C[a] (t/ha)	C:N Ratio
Coniferous forest	1.5–3	90
Deciduous forest	1.5–4	25–44
Temperate grassland	2–4	25–40
Legumes (medics and clovers)	1–2	10–20
Dung from dairy cows at a high stocking rate	1.7–2.3	20–25
Arable farming (cereal straw and stubble)	1–2	40–120
Chenin Blanc vines at 1680/ha		
leaf fall and prunings	1.2	22
Cabernet Sauvignon vines		
leaves	1	26–44
2-year-old wood	—	80

[a]For an approximate conversion to organic matter, multiply these figures by 2.5.
Source: Data from Williams and Biscay (1991), Mullins et al. (1992), and White (1997)

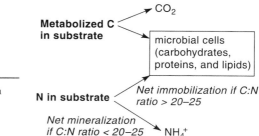

Figure 2.10   The influence of the C:N ratio of an organic substrate on the balance between net mineralization or immobilization of N.

2.3.1.2   Importance of the C:N Ratio

Grapevines, like all plant material, are composed of carbohydrates, proteins, and fats, plus smaller amounts of organic acids, lignin, pigments, waxes, and resins. The bulk of the material is carbohydrate, of which sugars (mainly sucrose) and starch are rapidly decomposed, whereas the structural molecules such as hemicellulose, cellulose, and especially lignin are less readily decomposed. Most fats, waxes, and resins are only slowly decomposed. Protein breakdown begins in the senescing leaves before they fall to the ground, as evidenced by the yellowing of older grapevine leaves when the berries reach full ripeness. Complete protein breakdown to amino acids is rapid once the leaves reach the soil. Microorganisms invade the litter and use some or all of the amino-N to synthesize their own proteins. Whether there is enough amino-N to meet the needs of the microorganisms depends on the C:N ratio of the substrate, as well as on the demands for growth of the decomposer organisms, as shown in figure 2.10. Any surplus N appears in the soil as NH_4^+ ions, which are available for plant growth.

The C:N ratio of fresh litter and dung from several sources is given in table 2.3. As the organic matter passes through successive cycles of decomposition, the C:N ratio becomes smaller. For example, the C:N ratio of *SOM* in well-drained soils of pH $\approx$ 7 is very close to 10. The ratio tends to be higher in poorly drained soils or in soils on which mor humus forms (section 1.3.3.1).

2.3.2   The Soil Biomass

Microorganisms are mainly responsible for the decomposition of complex organic molecules in litter and *SOM*; they are called "*decomposers*." However, in the decomposition process, there is much interdependence between the microorganisms and small animals, or the mesofauna. The digestive processes of the mesofauna are relatively inefficient, but they physically break residues into smaller and smaller pieces as they feed, creating a larger surface for the microorganisms to colonize. For this reason they are called "*reducers*." The intimate interaction among the reducers, the decomposers, and the residues they feed on is sometimes called a "food web." Collectively, the mass of organisms in a given quantity of soil is called the *soil biomass*.

2.3.2.1   The Decomposers

The decomposers consist of the *microflora* (members of the plant kingdom such as algae, bacteria, fungi, and actinomycetes) and *microfauna* (soil animals <0.2

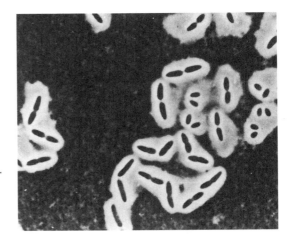

Figure 2.11   An example of soil bacteria, *Azotobacter* (White 1997). Reproduced with permission of Blackwell Science Ltd.

mm in length, such as protozoa). Within the microflora, we may distinguish two broad groups based on their mode of nutrition: (1) *heterotrophs*, including many species of bacteria and all the fungi, which require C in organic compounds as a substrate for growth, and (2) *autotrophs*, including the remaining bacteria and most algae, which synthesize cell components from the C of CO_2. The energy comes from sunlight (in the case of the photosynthetic bacteria and algae) or is chemical energy derived from the oxidation of inorganic compounds. Microorganisms differ in their requirement for molecular O_2. Accordingly, they can be *aerobes,* which require O_2 as the terminal acceptor for electrons generated in respiration, *facultative anaerobes*, which normally use O_2, but can adapt to oxygen-free conditions by using NO_3^- or other inorganic compounds as electron acceptors, and *obligate anaerobes*, which grow only in the absence of O_2 because O_2 is toxic to them.

   Bacteria and Actinomycetes. Bacteria are small—as small as 1 μm long and 0.2 μm wide. They are round or rod-shaped and live in water films around soil particles in all but the smallest pores (fig. 2.11). Under favorable conditions, bacteria multiply very rapidly so that their numbers in the soil can be enormous—between 1 and 4×10^9 organisms/g. The main limitation to their growth is the supply of substrate. Bacteria show almost limitless variety in their metabolism and ability to decompose diverse substrates.

   Actinomycetes are heterotrophs that form mycelial growths more delicate than the fungi (see next paragraph). This group includes such genera as *Streptomyces*, which produce antibiotics and can degrade resistant C compounds such as lignin.

   Fungi. The majority of fungi produce long filamentous hyphae, 1–10 μm in diameter, which may be segmented and/or branched. The network of hyphae, called a mycelium, develops fruiting bodies on which the spores formed are often highly colored. For example, members of the Basidiomycetes, or white-rot fungi, and spore formers such as *Penicillium* and *Aspergillus*, are often conspicuous on decaying wood and leaf litter in moist situations (fig. 2.12).

   Some fungi are *soil inhabitants*. These fungi are predominantly saprophytes, which means that they feed on dead organic matter in the soil. Fungi whose spores are normally deposited in the soil, but which only germinate and grow when living tissue of a suitable host plant is close by are called *soil invaders*. These fungi

Figure 2.12   An example of fungal mycelium and fruiting bodies. Photograph by the author. See color insert.

live as parasites and many are pathogenic. Downy mildew of grapevines is one such example. Of special benefit, however, are the *mycorrhizal fungi*, which live symbiotically in plant roots, deriving C compounds from the host and in turn supplying the host with mineral nutrients, especially P and Zn (section 4.7.3.2).

Soil fungal colonies are much less numerous than bacteria ($1–4 \times 10^5$ organisms/g). But because of their filamentous growth, their biomass is generally larger than the bacterial biomass. Fungi, excluding yeasts, are intolerant of anaerobic conditions. They grow better in acid soils (pH <5.5) and tolerate variations in soil moisture better than bacteria. Fungi are all heterotrophic and are most abundant in the litter layer and organically rich surface horizons of soil, where their ability to decompose lignin gives them a competitive advantage.

Algae. Algae can photosynthesize and therefore live at the soil surface. But many will grow heterotrophically in the absence of light, if simple organic substrates are provided. The two major subgroups are the Cyanobacteria, or blue-green algae, and the green algae. The "blue-greens," as exemplified by the genera *Nostoc* and *Anabaena*, are important because they can reduce atmospheric nitrogen (N_2) and synthesize amino acids for growth. Therefore, they can contribute a substantial amount of N to wet soils. Algae that occur in soils are much smaller than the aquatic or marine species and may number from 100,000 to 3×10^6 organisms/g.

Protozoa. Protozoa are the smallest of the soil animals, ranging from 5 to 40 μm in length. They live and move in water films in the soil. Nearly all protozoa prey on other small organisms such as bacteria, algae, fungi, and even nematodes. Protozoan biomass is comparable to that of earthworms, so they are important in controlling bacterial and fungal numbers in the soil, and hence they are important in nutrient cycling.

Soil enzymes. Soil organisms "digest" substrate molecules through the action of enzymes, which can act inside or outside the cell. But there are also soil enzymes that exist independently of living organisms. They avoid denaturation and degradation by being adsorbed on soil mineral or organic matter. The major types present are hydrolases, transferases, oxidases, reductases, and decarboxylases. One very common and stable soil enzyme is *urease*, which is important in the hydrolysis of urea (section 5.4.1.2).

2.3.2.2   *The Reducers*

By far the most important reducers are the *arthropods* and *annelids*. Nematodes and molluscs are also reducers.

Arthropods. This group includes the wood lice, mites and springtails, insects (adults and larvae), centipedes and millipedes. Many *mites* and *springtails* feed on plant residues and fungi in the litter, especially where thick mats build up under forests and undisturbed grassland. Their droppings appear as characteristic pellets in the litter layer (fig. 2.13). Predatory adult mites and springtails feed on other mites and nematodes, whereas the juveniles feed on bacteria and fungi. Mites can be a pest of grapevines.

Many *beetles* and *insect larvae* live in the soil, some feeding saprophytically, whereas others feed on living tissues and can be serious crop pests. Coprophagous (or dung-eating) beetles are important in breaking down the dung of large herbivores, especially in warmer climates. Their action greatly increases the rate at which such residues are mixed with the soil. *Termites* are also important in tropical and subtropical regions because they attack all forms of litter—tree trunks, branches, and leaves—under forests and especially in grasslands that receive seasonal rainfall (savannas). In cool temperate regions, various species of ant are of some importance in breaking down litter under pine forests and grassland. *Centipedes* are carnivorous, feeding on other small animals, whereas *millipedes* feed on vegetation, much of which is in the form of living roots, bulbs, and tubers.

Annelids. Earthworms are the most important group of annelids because of their size and physical activity. Under favorable conditions, they are more important in consuming leaf litter than all the other invertebrates put together. The

---

Figure 2.13   Fecal pellets of mites in a litter layer (White 1997). Reproduced with permission of Blackwell Science Ltd.

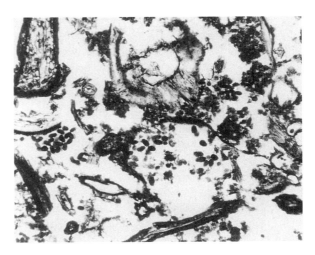

Figure 2.14   Earthworm casts on the soil surface. Photograph by the author. See color insert.

exceptions occur in the seasonal-rainfall tropics and subtropics (where termites are generally predominant) and in soils under coniferous forests and heath where mor humus forms.

Earthworms feed exclusively on dead organic matter. A few species, such as *Lumbricus rubellus* and *L. castaneous*, live mainly in the surface soil and litter layer, provided that temperature and moisture conditions are favorable. The majority of earthworms burrow more deeply into the soil, and in the course of feeding ingest large quantities of clay and silt-size particles. Consequently, organic and mineral matter is more uniformly mixed when deposited in the worm feces, which may appear as *casts* on the soil surface (fig. 2.14). In a study of several vineyards on contrasting soil types in the Barossa Valley of South Australia, the most abundant species found was *Aporrectodea rosea*, followed by *A. caliginosa*. The latter species is sensitive to dry soil conditions and was not found in the two sandiest soils (Buckerfield and Webster 1996). Other useful species are *A. trapezoides* and the two deep-burrowing species *A. longa* and *Lumbricus terrestris*.

As a result of earthworm activity, organic matter from the surface becomes incorporated throughout the upper part of the soil profile. Earthworms are most abundant under permanent grass swards because they are undisturbed and their food supply is plentiful. Their numbers may then exceed 250 per m^2, or 2.5 million per ha. The size of the population is regulated not only by the amount of suitable organic matter available, but also by soil temperature, moisture, and pH. For example, few species of earthworms are found in soils of pH less than 4.5: most species prefer neutral or calcareous soils. In hot, dry weather they burrow deeply into the soil and estivate.

Table 2.4 shows the range of earthworm biomass found under different management practices in a vineyard on an acid soil. Straw mulch provides extra food

Table 2.4   **Earthworm Biomass Under Different Management Practices**
              in a Barossa Valley Vineyard

Management Practice	Earthworm Biomass (kg/ha)
Mulch of grape marc (pomace)	220
Ryegrass cover crop, slashed and thrown into the rows	300
Bare soil, surface application of lime	340
Straw mulch	450
Straw mulch plus lime	700

Source: Data from Buckerfield and Webster (2001)

for the earthworms, conserves soil moisture, and reduces surface soil temperatures. When combined with lime to raise the soil pH, straw mulch encouraged the highest earthworm biomass.

Nematodes and molluscs. Next to protozoa, the threadlike *nematodes* are among the smallest of the soil fauna, ranging from less than 1 mm to a few mm in size. They are plentiful in soil and litter and may feed on dead organic matter or living fungi, bacteria, or other nematodes. Root-knot and root-lesion nematodes can be serious parasites of grapevine roots (section 7.3.3.2).

Other quite primitive soil animals are the *molluscs* (slugs and snails), many of which feed on living plants and are therefore pests. Some species feed on fungi and the feces of other animals. Because their biomass is only 200–300 kg/ha in most soils, the contribution of molluscs to the decomposition of organic matter is limited.

2.3.2.3   *Measuring the Soil Biomass*

The biomass of the large and small soil animals is easy to measure by direct methods in which they are isolated from the soil, counted, and weighed. This is not the case with the microorganisms, which are intimately mixed with the *SOM* and, being very small, are difficult to isolate and count or weigh. Methods for measuring the microorganisms collectively—the soil *microbial biomass*—are discussed in appendix 2. Estimates of microbial biomass C range from 500–2,000 kg/ha to 15-cm depth. Biomass C is only a small fraction of the total soil C, from about 1–2% in arable soils to 2–3% in grassland soils.

2.3.3   *Decomposition and Formation of Humus*

Humification is the process by which identifiable plant parts are physically and chemically altered to a black, amorphous material called *humus*. During humification, plant carbohydrates and proteins are decomposed and their microbial analogues are synthesized. Polymerization reactions involving aromatic compounds (unsaturated ring structures of C and H, of which the simplest is benzene C_6H_6) also occur. For example, simple hydroxy-phenols occurring naturally in plants, or produced from the breakdown of lignin and polyphenolic pigments, are oxidatively polymerized to form humic precursors. Nitrogen is incorporated into the polymers if amino acids combine with the phenols before polymerization. An im-

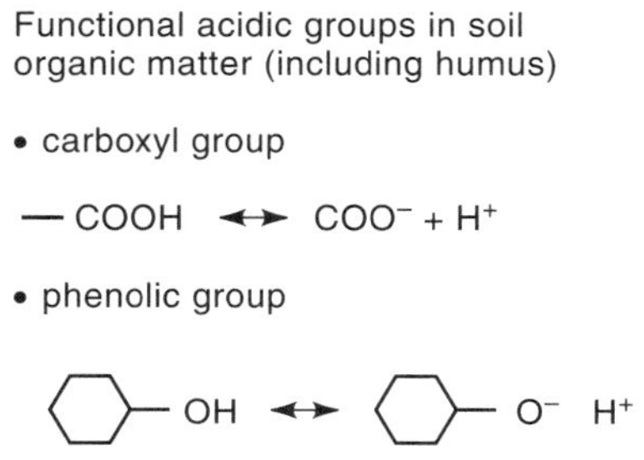

Functional acidic groups in soil organic matter (including humus)

• carboxyl group

— COOH ↔ COO⁻ + H⁺

$— COOH \leftrightarrow COO^- + H^+$

• phenolic group

Figure 2.15   Example of carboxyl and phenolic-OH groups in soil organic matter.

portant outcome of humification is that the number of acidic groups—carboxyl and phenolic groups (fig. 2.15)—increases per unit weight, thus contributing to the cation exchange capacity. This slow polymerization of phenolic compounds is analogous to the polymerization reactions, and "softening" of the tannins, that occur as a wine matures.

Because of the diversity of complex molecules formed during humification, identification of the chemical structure of humus has proved difficult. Much of the humus is strongly adsorbed to soil minerals, so ultrasonic vibration and strong solvents are needed to extract it. The treatment is severe, and it is possible that the compounds extracted are not the same as those in the soil originally. A widely used empirical procedure for extracting humic compounds in strong alkali followed by acid produces the chemical fractions *humic acid* (HA), *fulvic acid* (FA), and *humins*. More details are given in box 2.7.

| Box 2.7 | *Composition and Properties of Humic Fractions* |

The "core" structure of the HA and FA fractions is composed of aliphatic (a long hydrocarbon chain) and aromatic groups covalently bonded to form high molecular weight polymers, with much branching and folding. Polysaccharides and protein materials occupy the spaces in the folded and branched macromolecules, but they do not contribute more than about 20% of the total mass. These molecules may be covalently bonded or held by electrostatic attraction or hydrogen-bonding.

Approximately 35–45% of the HA fraction is aromatic, consisting mainly of single ring structures that are highly substituted, particularly with carboxyl (-COOH) and phenolic-OH groups of varying acid strength. Long aliphatic chains, also with carboxyl groups, link the aromatic structures or exist as separate side chains attached to the rings. Approximately 25% of the FA fraction is aromatic. FA molecules are smaller, more highly charged, and more polar than HA molecules. Repulsion of negative charges causes the FA molecules to be more linear than the randomly coiled HA molecules. FA molecules are soluble under acidic conditions and susceptible to leaching.

The steric arrangement of different functional groups (-COOH, phenolic-OH, and carbonyl C=O) facilitates the complexing of metal cations. The nonextractable (insoluble) humins are thought to be HA-type compounds that are strongly adsorbed, or precipitated, on mineral surfaces as metal salts or chelates.

2.3.4   *Other Properties of Humus*

2.3.4.1   *Cation Exchange Capacity*

Humus is an organic colloid of high specific surface area and high cation exchange capacity (*CEC*). Of the several functional groups containing O, those that dissociate H^+ ions—the carboxylic and phenolic groups—are the most important. The former have pK values between 3 and 5, whereas the latter begin to dissociate protons only at pH >7. The meaning of pK values and buffering is explained in appendix 3. Thus, the CEC of soil organic matter increases as the pH increases, and *SOM* has buffering capacity over a wide range of pH. The *CEC* is between 150 and 300 cmols charge (+) per kg dry matter, which is very significant in a soil with a low clay content. For example, if the organic matter content of a soil's A horizon is 5%, its contribution to the *CEC* of this horizon would be 7.5–15 cmols charge/kg soil. *CEC* is important for the retention of nutrient cations such as Ca^{2+}, Mg^{2+}, and K^+ (see chapter 4).

2.3.4.2   *Chelation*

Humic compounds form coordination complexes with metallic cations by displacing some of the water molecules from the cation's hydration shell. The stability of the resultant complex is enhanced through the "pincer effect" of the coordinating groups, giving rise to a *chelate* compound (fig. 2.16). Chelates formed with divalent and polyvalent cations are the most stable, with the stability constants decreasing in the order

$$Cu > Fe \approx Al > Mn \approx Co > Zn$$

Metal complexes formed with the HA fraction are largely immobile, but the more ephemeral Fe^{2+}-polyphenolic complexes are soluble and their leaching contributes to soil profile development in Podzols and Podzolic Soils (box 1.5).

2.3.5   *Factors Affecting the Rate of Organic Matter Decomposition*

2.3.5.1   *Type of Substrate*

Because various organic materials are decomposed in soil, we can expect a range of decomposition rates. The rate of substrate decomposition is assumed to follow simple first-order kinetics, according to the equation

$$\frac{dC}{dt} = -kC \tag{2.1}$$

dC/dt is the differential calculus expression for the rate at which the amount of substrate *C* changes with time *t*. The parameter *k* (the reciprocal of time) is the

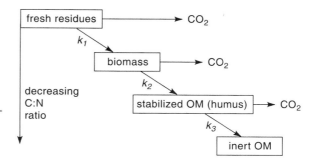

Figure 2.17   The "cascade effect" in organic matter decomposition.

first-order rate constant. When fresh residues are added to the soil, there is a "cascade effect" as the most easily decomposed parts are attacked first and progressively more resistant parts accumulate. Virtually all the C passes through the microbial biomass, which in turn dies and decomposes, adding to the more resistant remains. The process is illustrated in figure 2.17. The "stabilized" organic matter may consist of chemically resistant compounds, as in humus, or be physically protected, because of adsorption or its location in very small pores (see below). In the field, there is a wide spectrum of k values for different C substrates, as expressed by their half-lives $t_{0.5}$, where

$$t_{0.5} = \frac{0.693}{k} \tag{2.2}$$

These half-lives range from <1 month to >30 years, with the half-life for inert organic matter exceeding 50,000 years. However, for practical purposes, we can use average values of k in equation 2.1 to model the rate of *SOM* decomposition in individual soils.

2.3.5.2   *Soil Properties and Environmental Factors*

Soil moisture, O_2 supply, pH, and temperature affect the rate of organic matter decomposition. The first two factors tend to counteract one another because when soil moisture is high, a deficiency of O_2 may restrict decomposition. However, when the soil is dry, moisture but not O_2 will be the limiting factor. pH has little effect, except below 4 when the decomposition rate slows, as in the case of mor humus or cold, wet peaty soils. On the other hand, temperature has a marked effect, not only on plant growth and hence litter return, but also on litter decomposition through its effect on the rate of microbial respiration. For example, organic residues in soils of humid, hot regions (mean temperature 26°C) can decompose up to four times faster than in soils of humid, temperate regions (mean temperature 9°C).

Adsorption of C compounds by clays and sesquioxides generally slows down their rate of decomposition. For example, the accumulation of organic matter in soils derived from volcanic ash is attributed to the stabilization of HA and FA fractions through their adsorption on the clay minerals imogolite and allophane. Organic matter held in soil pores <1 μm diameter in clay soils (sometimes called "sterile pores") is also less accessible to microbial attack. Cultivation tends to break down soil structure so that organic matter in small pores is exposed to microor-

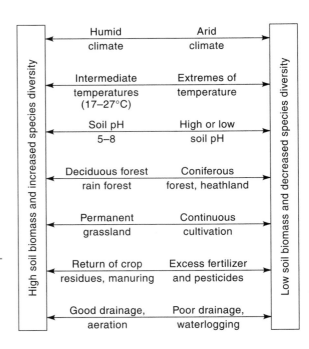

Figure 2.18   Dynamics of soil biomass changes (White 1997). Reproduced with permission of Blackwell Science Ltd.

ganisms and its decomposition rate is accelerated. Higher surface soil temperatures are also attained when the vegetative cover and protective litter layer are removed.

2.3.5.3   *Activity of the Soil Biomass*

The size and species diversity of the soil microbial population respond to environmental factors and soil properties, as described in preceding sections. Soil animals are also affected by types of litter, temperature variations, moisture, pH, and management practices, such as cultivation and the use of agricultural chemicals.

Environmental factors and soil management also influence the physiological activity of soil organisms. Striking effects occur, for example, when the soil is partially sterilized with toxic chemicals, such as ethylene dibromide. When favorable conditions are restored, rapid multiplication of the surviving organisms, feeding on the bodies of killed organisms, produces a flush of decomposition and a surge in the production of CO_2. Similar flushes of decomposition occur in soils subjected to air-drying followed by rewetting and in soils of cold temperate regions when they pass from a frozen to thawed state.

The dynamics of changes in soil biomass in response to variations in soil and environmental factors are summarized graphically in figure 2.18.

2.4   ***Summary Points***

The main points presented in this chapter on mineral and organic matter in soil follow.

■ The solid phase of soil, comprising ca. 50% by volume, consists of minerals and organic matter. The particle size distribution of the mineral matter is measured on the fine earth fraction (<2-mm sieve fraction). *Gravel* (2–600 mm) and *boulders* (>600 mm) are larger than the fine earth.

▓ The proportion of sand, silt, and clay-size particles in the fine earth determines the *soil's texture*. Texture is an important property that affects the soil's response to water (soil stickiness, consistence, and permeability), temperature, and cultivation (soil tilth and compaction). Texture can be assessed in the field.

▓ The clay-size fraction is divided into the *crystalline clay minerals* (layer-lattice minerals or phyllosilicates) and the *accessory minerals*, comprising various Fe and Al oxides (sesquioxides), Mn and Ti oxides, SiO_2, and calcite. The basic crystal structure of the phyllosilicates consists of sheets of Si in tetrahedral coordination with O $[SiO_2]_n$ and sheets of Al in octahedral coordination with OH $[Al_2(OH)_6]_n$. Crystal layers formed by the sharing of O atoms between contiguous silica and alumina sheets give rise to different minerals, depending on the Si:Al mole ratios: ≤ 1 (*imogolite* and the *allophanes*), 1:1 (*kaolinites*), and 2:1 (*illites, hydrous micas, vermiculites,* and *smectites*). Isomorphous substitution ($Al^{3+} \rightarrow Si^{4+}$ and Fe^{2+}, $Mg^{2+} \rightarrow Al^{3+}$) in the lattice results in an overall net permanent negative charge.

▓ The edge faces of the clay crystals, especially kaolinite, and the surfaces of Fe and Al oxides (e.g., *goethite* and *gibbsite*) bear variable charges depending on the association or dissociation of H^+ ions at exposed O and OH groups. The surfaces are positively charged at low pH and negatively charged at high pH. These charges are therefore called pH-dependent.

▓ The cation and anion charges adsorbed (mols of charge per unit mass) measure a mineral particle's cation exchange capacity, *CEC*, and anion exchange capacity, *AEC*, respectively. *CEC* ranges from 3 to 150 cmols charge (+) per kg for clay minerals, and *AEC* is 30–50 cmols charge (−) per kg for sesquioxides. Imogolite and allophane have roughly comparable *CEC* and *AEC* values at pH 6–7.

▓ The smaller a particle's size, the larger is its specific surface area. Depending on the mineralogy of the clay, the specific surface area ranges from 5 to 1000 m^2/g. Clay particles (<2 μm) are much more effective than silt (2–50 μm) and sand (50–2000 μm) in adsorbing water, ions, and solute molecules. In expanding-lattice clays, the total surface area is the sum of the internal and external areas; this can be as high as 750 m^2/g for a fully dispersed Na-montmorillonite (a smectite clay).

▓ Inputs of C as plant litter, dead roots, animal remains, and excreta are substrate for a heterogeneous population of soil organisms. As the organic residues decompose, releasing a proportion of the C as CO_2 to the atmosphere, soil organic matter (*SOM*) accumulates in a humified state. Soil humus consists of resistant plant residues (lignins), as well as complex new organic compounds synthesized by soil microorganisms.

▓ The living organisms *(biomass)* consist of the *reducers*, invertebrates such as mites, springtails, insects, termites, earthworms, nematodes, and molluscs, and the *decomposers*, microorganisms (bacteria, actinomycetes, fungi, and algae) and microfauna (protozoa). The most important reducers are *earthworms* in temperate soils, especially under grassland, and *termites* in many soils of the tropics and subtropics. These organisms of diverse size and function comprise an interdependent "food web" in the soil.

▓ Annual rates of litter fall are 0.9–1.2 t C/ha in vineyards of low planting density, but up to 4 t C/ha in temperate forests. Dead roots and carbon substrates deposited in the rhizosphere also contribute to biomass substrate. Earthworm biomass is 200–1500 kg/ha (up to 900 kg/ha in vineyards), whereas the microbial biomass is 0.5–2 t C/ha.

■ *SOM* serves as a reserve of C and other essential elements for successive generations of organisms. Elements sequestered in the biomass are said to be *immobilized*. Any surplus released in inorganic form is *mineralized*. For N, a substrate C:N ratio >25 favors an increase in microbial biomass and *net immobilization*, whereas a C:N ratio <25 favours *net mineralization*.

■ Humus is conveniently fractionated into *fulvic acid* (FA), *humic acid* (HA), and a nonextractable residue of *humins*. The FA and HA fractions are made up of macromolecules with a "core" of substituted aromatic and aliphatic C groups. A high density of acidic carboxyl (-COOH) and phenolic-OH groups confers a high *CEC* (150–300 cmol of charge/kg), which is pH-dependent. These acidic groups also form stable complexes (*chelates*) with metallic cations.

■ The rate of decomposition of litter and *SOM* can be expressed in terms of the half-life of the constituents, which ranges from <1 month to >50,000 years. The rate of decomposition is modified by clay content, moisture, temperature, and pH.

3 *The Vine Root Habitat*

3.1 *Root Growth and Soil Structure*

In the deep gravelly soils of the Bordeaux region, Seguin (1972) found vine roots at a depth of 6 m. Woody "framework roots" tend to be at least 30–35 cm below the surface and do not increase in number after the third year from planting (Richards 1983). Nevertheless, smaller diameter "extension roots" continue to grow horizontally and vertically from the main framework. They may extend laterally several meters from the trunk. These roots and finer lateral roots in the zone 10–60 cm deep provide the main absorbing surfaces for the vine. But in soils with a subsoil impediment to root growth, such as many of the duplex soils in southeast Australia (section 1.3.2.1), less than 5% of vine roots may penetrate below 60 cm (Pudney et al. 2001). Nor do vines root deeply in vineyards where irrigation supplies much of the vine's water in summer.

Plant roots and associated mycorrhizae (section 4.7.3.2) help to create soil structure. A desirable soil structure for vines provides optimal water and oxygen availability, which are fundamental for the growth of roots and soil organisms. The structure should be porous and not hard for roots to penetrate, allow ready exchange of gases and the flow of water, resist erosion, be workable over a range of soil water contents, allowing the seedlings of cover crops in vineyards to emerge, and be able to bear the weight of tractors and harvesting machinery with a minimum of compaction. The quality of soil structure and its maintenance in vineyards are discussed further in chapter 7.

3.1.1 *The Creation and Stabilization of Soil Structure*

We might expect the soil particles described in chapter 2 simply to pack down, as happens in a heap of unconsolidated sand at a building site. However, if the sand is mixed with cement and water, and used with bricks, we can construct a building—a solid framework of floors, walls, and ceilings. This structure has internal spaces of different sizes that permit all kinds of human activities. So it is with soil. Vital forces associated with the growth of plants, animals, and microorganisms, and physical forces associated with the change in state of water and its movement act on loose soil particles. The particles are arranged to form larger units called *aggregates*.

The formation of aggregates (also called *peds*) of various sizes and shapes determines the porosity of the soil. Both features—peds and pores—have been used to define soil structure, which in the broadest sense is denoted by the size, shape, and arrangement of soil particles and aggregates; the size, shape, and arrangement of the pores separating the particles and aggregates; and the combinations of aggregates and pores to form different structures.

For optimum plant growth, the soil structure should have a predominance of aggregates from 5 to 10 mm in diameter. The physical, chemical, and biotic forces involved in structure formation and stabilization include the following:

- Swelling pressures generated through the osmotic effects of exchangeable cations adsorbed on clay surfaces (box 3.1).
- Forces associated with the change in state of soil water. For example, tension forces develop as water evaporates from soil pores, pulling particles together. Unlike any other liquid, water expands on cooling from 4 to 0°C. Because water in the largest pores freezes first, liquid water is drawn from the smaller pores thereby subjecting them to shrinkage forces. But the expansion of the water on freezing, and the buildup of ice lenses in large pores, subject these regions to intense disruptive forces. For example, the action of frost in cold climates breaks down the massive clods left on the surface of a clay soil after autumn cultivation, producing a mellow "frost tilth" of numerous small granules (fig. 3.1).

Figure 3.1   Friable tilth created by frost in cultivated soil. Photograph by the author. See color insert.

- Burrowing invertebrates, which reorganize soil particles. The most important of these is the earthworm, which grinds and mixes organic matter with clay and silt particles in its gut, adding extra calcium in the process to form a *cast* (see fig. 2.14). Casts develop greater strength as they dry.
- The physical binding of soil by fine roots and fungal mycelia, including mycorrhizae. Particularly important in the stabilization of aggregates, this binding is one of the beneficial effects of having a cover crop, including perennial grasses, in the inter-rows of a vineyard.
- The activity of soil microorganisms, which can induce a change in metal valency leading to the solution and subsequent reprecipitation of Mn and Fe oxides (section 2.2.4.3). Fe oxides are usually positively charged and, being attracted to negatively charged clay minerals, are particularly effective cementing agents within soil aggregates.

3.1.2　*Describing Soil Structure*

Most soils have a structure that is readily observed in the field: this is the *macrostructure*. The USDA Soil Survey Manual (Soil Survey Division Staff 1993) presents a widely used system for describing and classifying soil macrostructure. This system, first published in 1951, is the basis for similar systems developed in other countries, such as in Australia (McDonald et al. 1990). However, much of the fine detail of particle interaction and aggregation is in the *microstructure*, which can be observed only using some form of magnification, ranging from a hand lens ($\times$10) in the field to a light microscope ($\times$50) or an electron microscope ($\times$10,000) in the laboratory. Descriptive terms for soil macrostructure are discussed in section 3.2.1.

Soil structure should be described from a freshly exposed soil profile or a large undisturbed core. Samples from augers (see chapter 8) are unsuitable because of the disturbance, as are those from old road cuttings and gullies that have been exposed to the weather for some time.

3.2　*Soil Aggregation*

True aggregates are recognizable because they are separated by pores and natural planes of weakness. They persist through cycles of wetting and drying, as distinct from the less permanent clods formed by digging or cultivation. A *concretion* is formed when localized accumulations of an insoluble compound, such as Fe oxide, cement soil particles together or enclose them. *Nodules* are similarly formed, but lack the symmetry and concentric internal structure of concretions.

3.2.1　*Aggregate Type, Class, and Grade*

Aggregates are described in more detail according to their *type*, *class*, and *grade*. The main *types* of aggregate are described in table 3.1.

Aggregate *class* is based on size. Size classes range from <2 mm to >500 mm, as measured by the average smallest dimension. The classes are described as fine, medium, coarse, or very coarse. Refer to the specialist soil survey handbooks for details.

Table 3.1   Types of Aggregates and Their Characteristics

Type of Aggregate	Descriptive Features and Characteristics
Spheroidal  **Figure 3.2** (a) Surface soil crumb structure, class 0.5–1 cm.	Roughly equidimensional and bounded by curved or irregular surfaces that do not fit against adjacent aggregates. Those that are relatively dense and nonporous are called *granules*. Very porous aggregates, often of dark color throughout, are called *crumbs* (fig. 3.2a). Crumbs are formed by the interaction of soil organisms, roots, and mineral particles in topsoils under grasslands and deciduous forests. They contribute to a desirable structure.
Blocky  **Figure 3.2** (b) Subangular blocky aggregates, class 2–3 cm.	Equidimensional with curved or flat surfaces that are mirror images of surrounding aggregate faces. Those with flat faces and sharply angular vertices are called *angular*; those with a mixture of rounded and flat faces and more subdued vertices are called *subangular* (fig. 3.2b). Blocky aggregates are common in the subsoil of loams and clay loams where they often have distinctive colors, because the surfaces are coated with organic matter, clay, or sesquioxides. They may have live or dead roots attached, or show distinct root impressions. A blocky structure in the subsoil usually indicates a good rooting environment for vines.
Platy  **Figure 3.2** (c) Platy aggregates in the A2 horizon of a sodic soil, class 0.5–4 cm.	Aggregates with a limited vertical dimension. An aggregate that is thicker in the middle than at the edges is called *lenticular* (fig. 3.2c). Aggregates occur in an A horizon immediately above an impermeable B horizon or in the surface of soil compacted by machinery. Water moves laterally rather than vertically through a soil horizon with platy structure, which is undesirable in vineyards.
Prismatic  **Figure 3.2** (d) Prismatic aggregates in a clay subsoil, scale is 10 cm.	Aggregates bounded by flat vertical faces, with sharp edges and flat tops, for which the horizontal dimension is limited compared to the vertical. The brick-red aggregates in figure 3.2d indicate a well-drained soil, but large prismatic aggregates that occur in the B horizon of clay soils subject to frequent waterlogging indicate poor drainage, an undesirable condition in vineyards.
Columnar  **Figure 3.2** (e) Columnar aggregates in a sodic subsoil, 12–15 cm across the columns. Photographs by the author. See color insert.	The same as for prismatic, but with rounded caps (fig. 3.2e). Columnar aggregates are typical of the impermeable B horizons of Na-affected soils; the main pathway for water movement is between the columns. However, because Na-clays swell when wet (box 3.1), these pathways close and subsoil drainage is severely impeded, which is very undesirable in vineyards.

Table 3.2   *Grade of Aggregates and Their Associated Structure*

Grade of Aggregate and Associated Structure	Description
Structureless or apedal	No observable aggregation nor definite natural planes of weakness. Structure is *massive*, if coherent, and *single grain* if noncoherent.
Weakly developed or weak	Poorly formed, indistinct aggregates that are barely observable in situ. When disturbed, the soil breaks into a mixture of a few entire aggregates (<one-third), many broken aggregates, and much unaggregated material.
Moderately developed or moderate	Well-formed, durable, and distinct aggregates, but not sharply distinct in undisturbed soil. When disturbed, the soil breaks down into a mixture of many entire aggregates (one-third to two-thirds), some broken, and a little aggregated material.
Strongly developed or strong	Durable, distinct aggregates in undisturbed soil that adhere only weakly, or separate, when the soil is disturbed. Soil consists largely of entire aggregates (>two-thirds), with a few broken aggregates and little or no unaggregated material.

The *grade* of an aggregate expresses its distinctness and durability (table 3.2). The grade depends on the water content at sampling and the time for which the soil is exposed to air-drying. (This is why it is recommended that structure be described only on freshly exposed soil). Prolonged drying markedly increases the aggregate grade, particularly in clay soils and soils high in Fe oxides. As indicated in section 2.2.2.3, the effect of soil conditions on aggregate grade is expressed through the term *soil consistence* (see box 2.2). Not only do individual aggregates have strength, and resist deformation, but the bulk soil also has strength, which is reflected in its capacity to bear loads and resist shearing forces. This is relevant to ameliorative treatments, such as deep ripping to improve drainage in vineyards (chapter 7).

3.2.2   *Factors Affecting Aggregate Stability*

For an aggregate to be stable, the between-particle forces must be strong enough to prevent the particle separating as a result of disruptive forces, such as those due to the impact of raindrops or heavy machinery. Stability depends on the way in which sand, silt, clay, and organic matter are organized to form aggregates. To understand this process, we must distinguish between naturally formed microaggregates (<0.25 mm) and macroaggregates (>0.25 mm diameter).

Microaggregates. A widely accepted model of a microaggregate consists of "domains" of clay minerals interacting directly with organic polymers and bonding to the O surfaces of quartz grains through these polymers (fig. 3.3). The formation and properties of domains are discussed in box 3.1. Prerequisites for the stability of microaggregates are flocculated clay and the ability to resist disruption on wetting or when mechanically disturbed. Clay-organic matter interactions involving polyvalent cations (Fe^{3+}, Al^{3+}, or Ca^{2+}) and humic polymers, or poly-

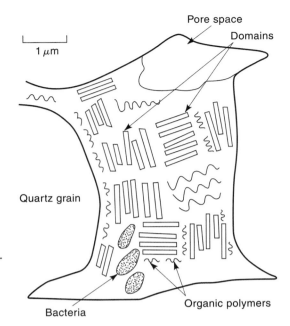

Figure 3.3   Diagram of a microaggregate
                    (White 1997). Reproduced with
                    permission of Blackwell Science
                    Ltd.

saccharides and clay surfaces, are also important. Inorganic compounds act as interparticle cements and stabilization agents. Examples of cementing agents are (1) $CaCO_3$ in calcareous soils, (2) sesquioxides as films between clay crystals and as discrete charged particles in many acid, highly weathered soils, and (3)SiO_2 at depth in highly weathered profiles, where a hardened or indurated layer may form.

Microaggregate stabilization is relatively insensitive to changes in soil management, except when management induces marked changes in the dominant cations on clays (e.g., Na^+ replacing Ca^{2+}), or changes in soil pH (acid → alkaline pH decreases the positive charges on kaolinite edge faces and sesquioxide surfaces), or physical disruption of aggregates by repeated cultivation that exposes "protected" organic matter to additional decomposition by soil organisms.

Macroaggregates. Figure 3.4 shows a model of a macroaggregate. The formation of a macroaggregate depends mainly on the stable cohesion of microaggregates in the larger structural unit. This stabilization is much more dependent on management than is stabilization of the microaggregates themselves, because it is achieved primarily through plant roots and fungal hyphae. Roots can penetrate pores >10 μm diameter, and fungal hyphae enter pores >1 μm. In so doing, they enmesh soil particles to form stable macroaggregates. Grass swards are especially effective because of the high density of fibrous roots in the surface soil.

Deposits of C compounds in the rhizosphere (section 2.3.1) provide the substrate for microorganisms that produce polysaccharide gums and bind soil particles together. These effects all depend on maintaining a high level of biotic activity—plants, animals (mainly earthworms), and microorganisms—in the soil. This activity is most important in soils of pH 5.5–7, which derive little benefit through aggregate stabilization by either sesquioxides or $CaCO_3$. Overall, macroag-

Box 3.1	*Formation and Properties of Domains*

Electron micrographs of thin sections of soil show crystals of montmorillonite, illite, or vermiculite stacked roughly parallel in the direction of the *c* axis (see fig. B2.4.1), to form clay domains. Flocculation of the clay crystals in a *face-to-face* array is necessary for domain formation. This kind of flocculation is favored when exchangeable cations of high charge (Al^{3+} or Ca^{2+}) are present, and the total salt concentration in the soil solution is high. For example, when Ca^{2+} is the dominant exchangeable cation, the basal spacing of the clay is at a minimum of 1.4 nm in a soil solution as concentrated as M $CaCl_2$. Spacing increases to no more than 1.9 nm as the solution concentration decreases to that of distilled water (fig. B3.1.1). Thus, each Ca-clay domain is a stable entity in water, showing limited swelling, because the expansion of the *diffuse double layer* (*DDL*) at the crystal surfaces is suppressed (section 4.5.2). The domains are only likely to be disrupted if more than about 15% of the moles of charge of Ca^{2+} are replaced by Na^+. Na^+ ions induce much greater swelling because these ions are less strongly attracted to the surface than Ca^{2+} ions, and the osmotic intake of water into the *DDL* is greater. If the Na-Ca clay domain is placed in distilled water, excessive swelling may lead to clay deflocculation (section 3.2.3).

In more highly weathered soils, where 1:1 clays such as kaolinite predominate, clay crystals are attracted in *edge-to-face* flocculation to form a "cardhouse" type of

Figure B3.1.1 Face-to-face flocculation with 2:1 clay mineral crystals and their expansion on diluting the soil solution (redrawn from White 1997).

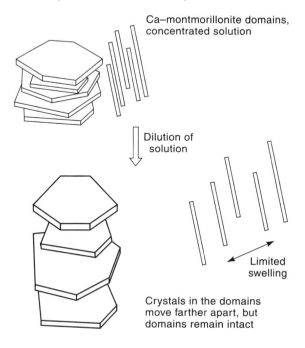

Ca–montmorillonite domains, concentrated solution

Dilution of solution

Limited swelling

Crystals in the domains move farther apart, but domains remain intact

(continued)

Box 3.1 *(continued)*

unit (fig. B3.1.2). Positively charged sesquioxide films on the flat surfaces also act as electrostatic bridges between clay crystals or form complexes with negatively charged organic polymers. These arrangements all contribute to the excellent structural stability of soils such as Krasnozems (Ferralsols in the World Soil Classification; see appendix 1). Krasnozems are important viticultural soils in parts of southeastern Australia.

Figure B3.1.2 Edge-to-face flocculation of kaolinite clay crystals (White 1997). Reproduced with permission of Blackwell Science Ltd.

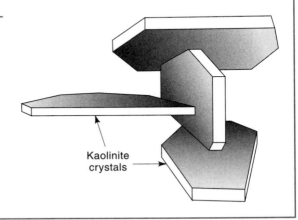

Figure 3.4   Diagram of a macroaggregate (White 1997). Reproduced with permission of Blackwell Science Ltd.

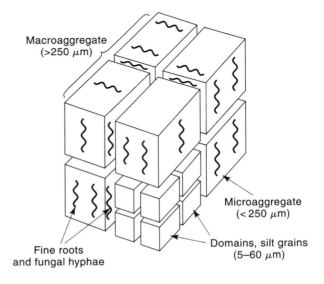

gregate stability is favored by substantial inputs of organic matter, biological activity, and minimal soil disturbance.

3.2.3   *Tests of Aggregate Stability*

The stability of aggregates is tested by their reaction to immersion in water. A long-established test measures *water stable aggregates* (Klute 1986). This method involves sieving, under water, an air-dry sample of soil on a vertical array of sieves of different mesh size. After a standard time, the mass of aggregates remaining on individual sieves is measured and expressed as a percentage of the total.

A useful field test of aggregate stability involves placing small air-dry aggregates (0.2–0.5 g) in water of very low salt concentration (e.g., rainwater). If the aggregate quickly collapses into subunits, which may be smaller aggregates or particles, it is said to *slake*. This indicates that the forces within an aggregate are not strong enough to withstand both the pressure of air entrapped as the aggregate wets up quickly and the pressure generated by clay swelling. The subunits produced by slaking may also be unstable if the swelling pressure in clay domains is strong enough to force individual clay particles apart, a process of *deflocculation* (or dispersion). An interpretation of soil structural stability based on the slaking and dispersive behavior of aggregates is given in box 3.2.

Flocculation, deflocculation, and swelling are clay-surface phenomena that are discussed further in chapter 4. However, deflocculation has several undesirable effects on soil structure: deflocculated clay blocks infiltration and drainage pores (chapter 7), translocated clay forms a dense clay B horizon, as in duplex soil profiles, and a hard surface crust forms when the deflocculated soil dries, impeding infiltration and seedling emergence.

3.2.4   *Soil Pans*

Translocation of colloidal inorganic and organic material from the A horizon into the subsoil is a major process in the creation of soil structure. In the changed physical and chemical environment of the subsoil, salts and oxides precipitate, clay particles and organic colloids flocculate, and deposits form on aggregate surfaces. Fe and Mn compounds that are reduced and dissolved under anaerobic conditions can be subsequently oxidized and deposited as oxidic coatings. These range in color from black MnO_2 minerals through to blue, blue-grey, yellow, and red Fe_2O_3 minerals, depending on the degree of oxidation and hydration of Fe. In dry environments, coatings of gypsum and calcite are found. Any of these coatings may build up to such an extent that bridges form between soil particles and the smallest aggregates, and a cemented layer develops in the soil. Such a layer or horizon is called a *pan*, which is described by its thickness (if <10 mm it is called "thin") and its main chemical constituent.

Iron pans commonly form at the top of the B horizon in duplex soils that are seasonally waterlogged. Over time and possibly as a result of climate change, these pans become dehydrated and fragment into a layer of rusty ironstone gravel, as shown in figure 3.5. Horizons that are cemented by silica deposits are called *duripans*: if massive, they may form a silcrete layer several meters thick, as found deep in the profiles of highly weathered soils in the southwest of Western Aus-

Box 3.2   *Slaking and Dispersion of Soil Aggregates in Water*

If dry aggregates do not slake in rainwater, the macroaggregate structural stability of the soil is good. Typically, aggregates from the A horizon of soils high in sesquioxides or formed on calcareous parent materials, and under permanent pasture, do not slake. Other soils may slake slowly or rapidly, depending on their organic matter content, sesquioxide content, and cultivation history.

Of the aggregates that slake, we differentiate types on the basis of whether they *disperse*, as shown in figure B3.2.1. Where there is no dispersion, a sample of dry aggregates is moistened and remolded into small balls, approximately 5 mm in diameter. These are placed in rainwater and observed for any dispersion. A description of the main types of aggregate behavior, identified by the test, is as follows.

Type 1. Complete dispersion, very poor microaggregate stability. The soil is likely to have no consistence when wet and to set hard when dry, and is also likely to erode easily. This soil is not recommended for vineyards.

Type 2. Partial dispersion, poor microaggregate stability; will probably crust and need to be protected by vegetative cover from erosion. Gypsum is required if the soil is to be cultivated and planted to vines.

Type 3. Complete to partial dispersion after remolding indicates moderate microaggregate stability. Soil structure will deteriorate under repeated cultivation, especially if cultivated when too wet; requires careful management in a vineyard.

Type 4. No dispersion after remolding indicates good microaggregate stability. The soil will not crust or erode very readily and is most suitable for vineyards.

Figure B3.2.1 Diagram of a modified Emerson dispersion test for soil aggregates (Cass 1999, after Emerson 1991). Reproduced from Soil Analysis: An Interpretation Manual (K. I. Peverill, L. A. Sparrow, and D. J. Reuter, 1999) with permission of CSIRO Publishing.

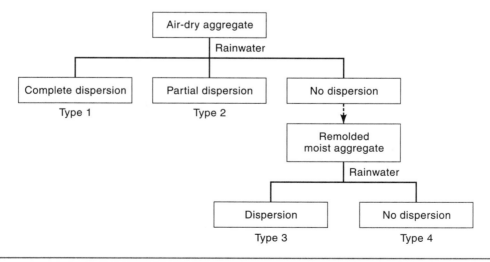

Figure 3.5   An ironstone gravel layer at the top of the B horizon in a duplex soil on sandstone in the Yarra Valley, Australia. Photograph by the author. See color insert.

tralia. Similarly, soil layers cemented by secondary $CaCO_3$ deposits are called calcrete. A *fragipan* (found in soils subjected to present or past freezing to depth) is not cemented, but the size distribution and packing of the soil particles produces a layer of a high bulk density. *Plow pans* are formed in cultivated soils at the maximum depth of cultivation (20–25 cm), as a result of the shearing force applied by the implements and compression by tractor wheels. These pans can be particularly bad in vineyards where the inter-rows are rotary-hoed (section 7.4.1).

3.3   *Soil Porosity*

Porosity is important because it determines the total amount of water the soil can hold. For a given soil volume, the porosity is defined as

$$\text{Porosity} = \frac{\text{Volume of pores}}{\text{Volume of soil}} \tag{3.1}$$

It follows that the pore volume is equal to the product of the porosity and the soil volume. This equation is most appropriate for nonswelling soils. In a soil with much expanding-lattice clay, such as montmorillonite, the total soil volume and the porosity may change appreciably as the water content changes.

3.3.1   *Bulk Density*

The simplest method of estimating porosity is from the soil's bulk density and particle density. *Particle density* ρ_p (rho p) is discussed in box 2.1. *Bulk density*

Box 3.3   *Measuring Soil Bulk Density*

BD is measured on intact soil cores, which are collected by driving steel or brass rings, usually 6 cm in diameter by 7.5 cm deep, into the soil. The rings should be thin walled (wall thickness <1/15 of the ring diameter) to avoid compaction. The soil-filled ring is dug out carefully, the soil trimmed level with the top and bottom, and the whole core wrapped in foil to prevent soil loss in transit to the laboratory. The mass of oven-dry (o.d.) soil is measured and the volume calculated from the internal dimensions of the ring. If the soil shrinks on drying, the dimensions of the dry soil core should be measured.

The *BD* is calculated from the equation

$$BD = \frac{\text{Mass of o.d. soil}}{\pi \times (\text{Radius})^2 \times \text{Length of core}} \qquad \text{(B3.3.1)}$$

Values of *BD* range from <1 g/cm^3 (or Mg/m^3) for soils high in organic matter to 1.0–1.4 for well-aggregated loamy soils, and to 1.2–1.8 g/cm^3 for sands and compacted B horizons in clay soils. *BD* values $\geq$ 1.6 g/cm^3 are generally undesirable in vineyard soils.

(*BD*) is defined as the mass of oven-dry soil per unit volume, and it depends on the soil's structure and the densities of the soil particles (clay, organic matter, quartz grains, etc.). A technique for measuring soil *BD* is described in box 3.3.

The equation for calculating porosity can be written as

$$\text{Porosity} = 1 - \frac{\text{Bulk density, } BD}{\text{Particle density, } \rho_p} \qquad (3.2)$$

The derivation of equation 3.2, and simple calculations using this equation, are given in box 3.4.

3.3.2   *Water-filled Porosity*

Soil pores are partly or wholly occupied by water. Normally, water occupies less than the total porosity, and the soil is said to be *unsaturated*. However, during a period of prolonged rain, the soil may temporarily become *saturated* when all the pore space fills with water. This condition can also occur when the water table rises toward the soil surface (section 6.1.1).

Soil wetness is assessed according to the amount of water held per unit mass or unit volume of soil:

- The *gravimetric water content* (θ_g or theta g) is measured by drying the soil to a constant weight at a temperature of 105°C. θ_g is then calculated from

$$\theta_g = \frac{\text{Mass of water}}{\text{Mass of oven-dry soil}} \qquad (3.3)$$

θ_g is often expressed as a percentage by multiplying by 100.

| Box 3.4 | *Useful Calculations Involving Soil Bulk Density* |

We may expand equation 3.1 to write

$$\text{Porosity} = \frac{\text{Volume of soil} - \text{Volume of soil solids}}{\text{Volume of soil}} \qquad \text{(B3.4.1)}$$

$$= 1 - \left(\frac{\text{Mass of o.d. soil}}{\text{Volume of soil}} \times \frac{\text{Volume of soil solids}}{\text{Mass of o.d. soil}} \right) \qquad \text{(B3.4.2)}$$

$$= 1 - \frac{BD}{\rho_p} \qquad \text{(B3.4.3)}$$

Using equation B3.4.3, we can show, for example, that a soil of *BD* 1.33 Mg/m^3 and particle density 2.65 Mg/m^3 has a porosity of 0.5 m^3/m^3 or 50%.

Laboratory soil analyses are normally reported on a unit mass basis (e.g., "available P" in μg P/g soil). However, such measurements often need to be related to the field, where units such as kg P/ha (or lb/ac) are more appropriate. Implicit in this field unit is a volume of soil, usually measured to a depth of 15 cm (6 in.), which approximates the cultivation depth. Laboratory measurements per unit mass can be translated to a field basis by calculating the mass of soil in 1 ha to a depth of 0.15 m, using the soil *BD*.

For example, for a soil of *BD* = 1.33 Mg/m^3, the mass *M* of o.d. soil per ha-0.15 m is

$$M \text{ (ha-0.15 m)} = 1.33 \times 0.15 \times 10^4 = 1995 \text{ Mg/ha} \qquad \text{(B3.4.4)}$$

The use of this relationship in equation B3.4.4 is demonstrated in subsequent chapters. The conversion to American units is given in appendix 15.

- The *volumetric water content* (θ_v or theta v) is given by

$$\theta_v = \frac{\text{Volume of water}}{\text{Volume of soil}} \qquad (3.4)$$

Box 3.5 shows the relationship between θ_g and θ_v.

3.3.3 *Air-filled Porosity*

Air fills the soil pore space not occupied by water. The air-filled porosity ϵ (epsilon) is defined by

$$\epsilon = \frac{\text{Volume of soil air}}{\text{Volume of soil}} \qquad (3.5)$$

from which it follows that

$$\epsilon = \text{Porosity} - \theta \qquad (3.6)$$

When a soil drains, after being thoroughly wet by rain or irrigation, the largest pores empty quite rapidly, usually within 2 days, after which the drainage rate

Box 3.5   *Useful Calculations Involving Soil Water Content*

The units of θ_v are commonly m³ water/m³ soil. Thus, the concentration C_l of a solute in the soil solution can be expressed on a soil volume basis (C_{soil}) by multiplying by θ_v:

$$C_{soil} = C_l \times \theta_v \qquad (B3.5.1)$$

θ_g and θ_v are related through the soil's *BD* by the equation

$$\theta_v = \theta_g \times BD \qquad (B3.5.2)$$

because 1 Mg water occupies 1 m³ at normal temperatures.

Equipment used to measure soil water content in the field (see box 6.9) gives a value for θ_v. Throughout this book, soil water content is expressed as a volumetric water content, using the symbol θ. Values of θ are a direct measure of the *equivalent depth* of water per unit area of soil surface, which is useful for comparison with rainfall, evaporation, or the depth of irrigation water applied. For example, assume an average θ value of 0.25 m³/m³ for a soil profile. This may be written as

$$\theta = \frac{(0.25 \text{ m} \times 1 \text{ m}^2 \text{ surface}) \text{ of water}}{(1 \text{ m} \times 1 \text{ m}^2 \text{ surface}) \text{ of soil}} \qquad (B3.5.3)$$

Conceptually, this is equivalent to a depth of 0.25 m (250 mm) of water per m depth of soil (fig. B3.5.1). The general rule is that $\theta \times 1000$ gives the equivalent depth of water (mm) in 1 meter depth of soil. If the soil depth of interest z is different from 1 m, the equivalent depth d (mm) of stored water is given by

$$d = \theta \times 1000 \times z \text{ (in meters)} \qquad (B3.5.4)$$

Figure B3.5.1 Diagram of a 1-m³ soil cube filled with water to a depth of 250 mm.

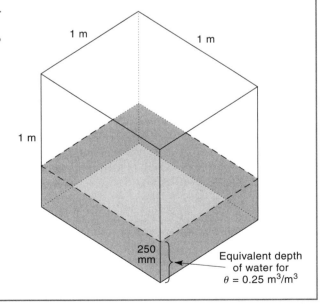

| Box 3.6 | *Calculating Air Capacities for Vineyard Soils* |

The air capacity is a good indicator of soil aeration for vines (section 3.4). Many compacted soils have low total porosity, poor drainage, and a low air capacity. For example, suppose the *BD* of a compacted soil is 1.6 Mg/m^3. From equation B3.4.3, this corresponds to a porosity of 0.40 m^3/m^3. If the *FC* value is 0.35 m^3/m^3, we have

$$C_a = 0.40 - 0.35 = 0.05 \text{ m}^3/\text{m}^3 \qquad \text{(B3.6.1)}$$

An air capacity of 0.05 m^3/m^3 or 5% is too low for adequate soil aeration for vines. However, suppose we have a well-structured loamy soil of *BD* = 1.2 Mg/m^3. The porosity will be 0.55 m^3/m^3, and if the *FC* value is 0.40 m^3/m^3, we have

$$C_a = 0.55 - 0.40 \text{ m}^3/\text{m}^3 = 0.15 \text{ m}^3/\text{m}^3 \qquad \text{(B3.6.2)}$$

An air capacity of 0.15 m^3/m^3 or 15% should provide good soil aeration and healthy root activity. The minimum air capacity for adequate aeration is 10% (see box 3.7).

---

slows. The soil water content θ at this point (in the absence of evaporation) defines the *field capacity* (*FC*). The pores drained at field capacity are sometimes called *macropores*, and correspond to the pores and cracks between the larger soil aggregates. Macropores are mainly created by soil fauna and roots (collectively called *biopores*) and by the shrinkage on drying of large aggregates in clay soils. The smaller pores within aggregates that remain water-filled at the *FC* are called *micropores*. The distinction between macropores and micropores is useful for defining the soil's *air capacity*, (*C$_a$*), which is obtained from equation 3.6 by substituting the value for θ at the *FC*. The calculation of air capacity is illustrated in box 3.6.

The subdivision of porosity into macropores and micropores, although useful, is a simplification of reality because the distribution of pore sizes in soil is continuous from large to very small. This distribution can be estimated from the soil water retention curve (section 6.2.1). It is also important to understand the empirical relationship between pore size and function, as outlined in table 3.3.

| Table 3.3 | *Relationship Between Pore Size and Function in Soil* |

Pore Diameter (μm)	Associated Biotic Agent and Function
5000–500	Earthworm channels and main plant roots; pores for rapid drainage and aeration
500–30	Grass roots and small mesofauna; normally draining pores; aeration
30–10	Fine lateral grass roots and fungal hyphae; very slowly draining pores
30–0.2	Root hairs, fungal hyphae, and bacteria; "available water" storage
<0.2	Swell-shrink water in clays; residual or "nonavailable" water storage

Source: After Cass et al. (1993) and White (1997)

3.4   **Soil Respiration and Aeration**

Soil fauna and microorganisms live in the soil's pore space. Plant roots grow through the pore space and also create pores as they grow. Living organisms and roots respire. Normally, respiration is *aerobic*—that is, O_2 is consumed and CO_2 and water are released as complex organic substrates are oxidized, as in the following example:

$$C_6H_{12}O_6(glucose) + 6O_2 \leftrightarrow 6CO_2 + 6H_2O \tag{3.7}$$

If the concentration of O_2 in the soil air falls to a low value ($<1/60$ in the air above the soil), aerobic respiration will cease. The roots of plants other than those adapted to waterlogged conditions, such as some sedges and rushes, go brown and die. The strictly aerobic soil organisms are replaced by facultative and obligate anaerobes (section 2.3.2.1), a change that is generally undesirable for soil health. Thus, the exchange of O_2 and CO_2 between the soil air and the atmosphere is most important to maintain a balanced gas composition in the soil. This process of gas exchange or aeration depends on the soil structure, and particularly on the ratio of macropores to micropores.

The proportions in which the major gases N_2, O_2, and CO_2 occur in the atmosphere are equivalent to partial pressures of 79.0, 21.3, and 0.0367 kPa, respectively. A dynamic equilibrium is established between the atmosphere and the soil air that depends on the soil's respiration rate and the resistance to gas movement through the soil. A soil in which the gases can move rapidly through the macropores has an O_2 partial pressure of $\sim$20 kPa and a CO_2 partial pressure in the range of 0.1–1 kPa. However, if the macropores fill with water and the soil becomes waterlogged, the soil becomes increasingly O_2 deficient or *anoxic*. In this case, the O_2 partial pressure may drop to zero and that of CO_2 may rise well above 1 kPa. If soil becomes completely anoxic and remains so for some time as a result of anaerobic respiration, the gas phase may consist almost entirely of CO_2 and CH_4 in roughly equal proportions. The chemical and biological changes that occur as a soil becomes progressively more anaerobic, some of which are undesirable, are described in section 5.6.

3.4.1   *The Mechanism of Gas Exchange*

Profiles of O_2 partial pressure in a poorly drained silty clay and a freely drained sandy loam, in both winter (wet soil) and summer (dry soil), are shown in figure 3.6. These contrasting profiles reflect the fact that the main mechanism of gas exchange between the soil and the atmosphere is gas diffusion. The rate of diffusion depends on the intrinsic diffusion coefficient of the gas molecules, the nature of the pathway for diffusion, and the gradient in gas concentration. These factors are discussed in more detail in box 3.7. An adequate rate of gas diffusion is also important for the effective fumigation of vineyard soils to control nematodes (chapter 7).

3.4.2   *Soil Management and Aeration*

Management practices that change the proportion of large, continuous pores in a soil affect aeration. Degradation of soil structure through continuous cultivation, for example, may not necessarily result in much change in total porosity, but it is

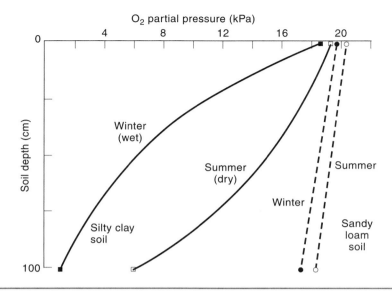

Figure 3.6   Depth profiles of O_2 partial pressure in contrasting soils during a dry summer and wet winter. Solid symbols, winter; open symbols, summer (redrawn from White 1997).

Box 3.7   *Factors Controlling the Rate of Gas Diffusion in Soil*

Diffusion is a molecular process whereby individual ions or molecules move in response to gradients in their concentrations between one point and another. Gases in the air diffuse relatively quickly; for example, the diffusion coefficients of O_2 and CO_2 in bulk air are about 0.2 cm²/sec at normal temperatures. But the diffusion coefficients of these gases dissolved in water are 10,000 times smaller. This difference has a marked effect on the rate of gas diffusion through gas-filled pore space, compared to water-filled pore space in soil. The rate of gas diffusion or *diffusive flux* J_g is the volume of gas diffusing across a unit area perpendicular to the direction of movement in a unit of time. (Because 1 mole of a gas at 0°C and 1 atmosphere's pressure occupies 22.4 L, gas quantities and concentrations are usually expressed in volume units). The equation for J_g is

$$J_g = -D \frac{dC}{dx} \tag{B3.7.1}$$

The term dC/dx is a differential calculus expression for the change in gas concentration C per unit distance x in the direction of diffusion. D is the gas diffusion coefficient, and when C is expressed in units of mL/mL and x is in cm, D is in cm²/sec. However, even in the air-filled pores, D values for O_2 and CO_2 in the soil are always less than their values (D_o) in bulk air, because of the restricted volume of the pore space and the tortuosity of the diffusion pathway. The effective diffusion coefficient in soil, D_e, is given by the equation

$$D_e = f\epsilon D_0 \tag{B3.7.2}$$

(continued)

Box 3.7 *(continued)*

where ϵ is the air-filled porosity and f is an impedance factor. The value of f, which is between 0 and 1, changes with ϵ in a complex way that depends on a soil's structure, texture, and water content. As water fills the small pores, the pathway for diffusion in the air becomes more tortuous and f decreases. Once the macropores between aggregates begin to fill with water, f decreases rapidly as ϵ decreases, falling close to zero at $\epsilon \approx 0.1$ m^3/m^3. This is why the lower limit for an acceptable air capacity in many soils is 0.1 m^3/m^3 or 10% (box 3.6). At $\epsilon <$ 0.1 m^3/m^3, many of the remaining air-filled pores are discontinuous or "dead end."

usually associated with a disproportionately large decrease in the macroporosity. Consequently, although the value of the impedance factor f may be as high as 0.6–0.7 in a well-structured soil, it may fall as low as 0.1 in an overcultivated, degraded soil at the same water content.

Even in a well-structured soil with adequate air capacity, the interiors of large aggregates may become anoxic under wet conditions (fig. 3.7) because O_2 has a low solubility in water. Although the O_2 partial pressure in the air-filled macropores may be 20 kPa, because of its low solubility and very slow rate of diffusion in water, the partial pressure of O_2 within aggregates more than 3–4 cm in diameter may be close to zero. CO_2, on the other hand, is relatively soluble in water (27 times more soluble than O_2 at the same temperature and pressure). When the O_2 pressure falls to zero in the center of an aggregate, the CO_2 partial pressure due to respiration at the center will only rise to about 1 kPa. The critical aggregate size to cause such anoxic conditions depends on the soil respiration rate, which, in addition to depending on the supply of C substrate, is highly dependent on soil temperature.

Figure 3.7   Anoxic zones within large wet aggregates in a structured soil (redrawn from White 1997).

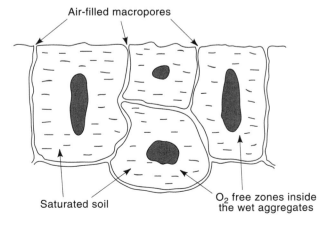

Air-filled macropores

Saturated soil

O_2 free zones inside the wet aggregates

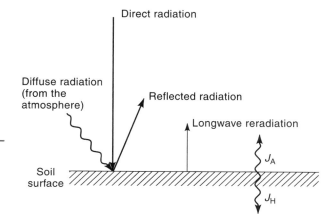

Figure 3.8　Radiation energy balance and heat fluxes at the soil surface (redrawn from Linacre 1992).

3.5　*Soil Temperature*

3.5.1　*Energy Fluxes and Temperature*

The energy balance at the soil surface primarily controls soil temperature. The main input of energy is solar radiation of which a fraction is directly reflected, depending on the reflectance or *albedo* of the surface. Some of the shortwave radiation that is absorbed by the soil surface and vegetation is reradiated to the atmosphere as longwave radiation. Of the remainder, called the *net solar radiation R_N*, the greatest part is dissipated by evaporation of water (section 6.4.2). The balance of the absorbed radiation is partitioned between the transfer of sensible heat to the air immediately above the ground (J_A) and the transfer of heat deeper into the soil (the soil heat flux J_H). A summary of the energy fluxes at the surface is shown in figure 3.8.

The size of J_H determines the effect of radiation on soil temperature. The value of R_N, and hence J_H, varies with the surface albedo, which depends on these factors: the nature and color of the surface, the type of vegetation, the latitude, and the slope of the land in relation to the sun. Representative values of albedo for different surfaces are given in table 3.4.

Table 3.4　***Representative Values of Albedo for Different Surfaces***

Type of Surface	Range of Albedo (%)	Median Value (%)
Tall grass (2 m) to short grass (0.3 m)	21–25	23
Field crops	15–24	20
Forests	10–18	15
Wet soil to dry soil	11–18	14
Water	4–13	7

Source: After Linacre (1992)

3.5.2   ***Factors Affecting Soil Temperature***

Color affects the albedo of a soil surface. For example, black and dark red surfaces reflect less radiation and therefore warm up more than light-colored surfaces. But at night, dark surfaces radiate energy (by longwave radiation) faster than light-colored surfaces. This radiation loss adds to the transfer of sensible heat by the process of convection. Bare soil surfaces with no vegetative cover lose heat rapidly at night, especially if there is no low cloud cover. However, the vine canopy traps heat emanating from the soil after dark and reduces heat loss at night, so that the surface soil temperature and near-surface air temperature remain higher than otherwise expected. This is important for frost avoidance.

Because of their low thermal conductivity, *mulches* of organic material such as straw or bark pieces reduce J_H (either downward during the day or upward at night). The presence of a mulch therefore moderates the temperature fluctuations between day and night (section 3.5.3), which can be particularly important in warm climate vineyards where high temperatures in the vine rows promote high soil evaporation rates. Lower daytime soil temperatures also favor earthworm activity and slower rates of organic matter decomposition. However, the insulating effect of a mulch may increase the chance of early morning frost in vineyards where this is a risk. Further, because both cover crops and mulches have a higher albedo than bare soil, the energy absorbed by the soil, and hence soil warming, is less under these covers, thereby increasing the risk of frost.

The effect of a given value of J_H on soil temperature depends on the soil's heat capacity per unit volume (C_b), and its thermal conductivity. Water has a much higher C_b than dry soil, so wet soil must absorb much more heat than dry soil to change the temperature by 1°C. In vineyards in cool climates, soil wetness can therefore be an important factor determining when root growth will occur. Vine roots start to grow at temperatures around 6°C (43°F), but they have an optimum close to 30°C (86°F).

3.5.3   ***Diurnal and Annual Temperature Changes***

The change in the direction of soil heat flux downward during the day to outward at night causes the soil temperature to change in a sinusoidal way over a 24-hour period. This is called the diurnal temperature variation. The longer term variation in heat flux between summer and winter causes a similar sinusoidal variation of much longer period—the annual temperature variation. These diurnal and annual variations in temperature T can be described by the equation

$$T = T_A + A \sin \left(\frac{2\pi t}{\tau} \right) \tag{3.8}$$

where t is time, T_A is the average temperature (daily or annual), τ (tau) is the period of the sine wave (either 1 day or 1 year), and A is the size of the temperature change.

An example of diurnal variation in soil temperature is shown in figure 3.9. The size of the variation is greatest at the surface and decreases with depth. Within a 24-hour period, either the maximum or minimum temperature is attained at a progressively later time as the depth increases. In addition to heat being transferred in a vertical direction across the soil surface, heat is transferred by lateral

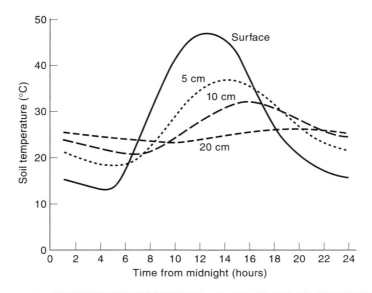

Figure 3.9   Diurnal variation in temperature at different depths in the soil (White 1997). Reproduced with permission of Blackwell Science Ltd.

air movement due to larger scale convection processes induced by the topography or by the presence of large water bodies. This is an important determinant of frost susceptibility in vineyards, which is discussed further in chapter 8.

3.5.4   *Soil Temperature and Respiration*

Soil respiration is a combination of root respiration and respiration by soil organisms. The rate of soil respiration determines the demand for O_2 in the soil, especially inside aggregates, and hence influences the rate of gas diffusion (box 3.7). The respiration rate (R) is normally measured as the rate of O_2 consumption per unit volume of soil in a unit of time. For aerobic respiration, 1 mole of CO_2 is produced per mole of O_2 consumed, so R can also be measured by the rate of CO_2 production under these conditions. The effect of temperature on the respiration rate is expressed by the equation

$$R = R_0 Q^{7/10} \tag{3.9}$$

where R and R_o are the respiration rates, respectively, at ambient temperature T and at a temperature of minimum respiration rate (0°C). The term Q defines the relative increase in R for a 10°C rise in temperature, called the *Q-10 factor*, which lies between 2 and 3. This applies up to the optimum temperature, generally between 30 and 35°C, above which enzyme function deteriorates.

In regions of cool climate, temperature change is the cause of a large seasonal variation in the soil respiration rate, which follows a sinusoidal trend as described by equation 3.8. In such regions, the seasonal maximum in R lags 1–2 months behind that of soil temperature. Superimposed on this seasonal pattern is the effect of weather, because within any given month the daily maximum in R may be up to twice the daily minimum, a fluctuation strongly correlated with surface soil temperatures. Also, there are variations in R in the surface soil (down to ca. 20 cm) caused by the diurnal rise and fall of soil temperature.

3.6   *Summary Points*

This chapter explores the influence of soil structure, temperature and biotic activity on the vine root habitat. The main points are summarized as follows:

◼ Physical, chemical, and biotic forces act on clay, silt, and sand particles, in intimate combination with organic materials, to form an organized arrangement of aggregates (*peds*) and intervening spaces (*pores*). This is the basis of *soil structure*.

◼ Good soil structure provides a favorable environment for the growth of roots and soil organisms. Good structure is particularly important for adequate aeration and drainage in vineyard soils.

◼ Plant roots, fungal hyphae, microorganisms, and burrowing animals such as earthworms are important biotic agents for creating and stabilizing soil structure.

◼ Aggregates are described according to their shape, size and grade. The grade of an aggregate depends on the soil's water content. *Soil consistence* is a practical measure of aggregate grade and gives an overall indication of soil strength (related to its friability and ease of penetration by roots).

◼ On a microscopic scale, clay crystals (such as illite, vermiculite, and montmorillonite) can stack in roughly parallel alignment, with much overlapping, to form coherent units called *domains*. These units are stable in water when Ca^{2+} is the dominant exchangeable cation, but become potentially unstable when 15% or more moles of charge of Ca^{2+} are replaced by Na^+.

◼ Clay domains, sesquioxides, and organic polyanions are drawn together by the attraction of positively and negatively charged surfaces, reinforced by interatomic forces at very close range (<1 nm) to form *microaggregates* (<250 μm in diameter). The combination of microaggregates into *macroaggregates* (>250 μm) is enhanced by microbial polysaccharide gums, which act like glue, and fungal hyphae and fine plant rootlets, which enmesh the aggregates. Good structure for plant growth depends on the existence of water-stable aggregates between 5 and 10 mm in diameter.

◼ In specific cases, sesquioxides, SiO_2, or $CaCO_3$ can act as interparticle cements. However, if the cementing agent is in excess, dense impermeable layers called *pans* may form. These pans are undesirable, as are plow pans formed by cultivating soil when it is too wet.

◼ The size, shape, and arrangement of aggregates determine the intervening volume of pores, or *soil porosity*. The porosity is defined as the volume of pores per unit volume of soil, and for nonswelling soils can be calculated from the equation

$$\text{Porosity} = 1 - \frac{\text{Bulk density}}{\text{Particle density}}$$

where *bulk density* is the mass of o.d. soil per unit volume, and the mean particle density is 2.65 Mg/m^3.

◼ Porosity normally ranges from 0.40 to 0.60 m^3/m^3 (40–60%). The pores can be occupied by water or air. Changes in soil wetness are measured by changes in *vol-*

umetric water content θ, and the *air-filled porosity* is defined as (Porosity $-$ θ). The minimum value of air-filled porosity for adequate soil aeration is 10%.

■ The drainage and aeration of soil depend mainly on the distribution of pore sizes. Pores and cracks between large aggregates that are drained at the field capacity are called *macropores*; they include many of the pores created by burrowing soil fauna and plant roots (the biopores). Pores that still hold water at the field capacity are called *micropores*.

■ Respiration by roots and soil organisms consumes O_2 and produces CO_2. In a well-aerated soil, the O_2 partial pressure in macropores is approximately the same as in the atmosphere (ca. 20 kPa), and that of CO_2 is 0.1–1 kPa. However, in the interior of aggregates 3–4 cm in diameter, which remain saturated, the O_2 partial pressure can fall to zero. The soil is then a mosaic of aerobic and anaerobic zones.

■ The development of anaerobic zones depends on the soil's respiration rate, which is markedly influenced by soil temperature. Soil temperature is determined by the net absorption of the sun's radiant energy, which is influenced by the surface cover and soil color.

■ Dark-colored soil surfaces absorb more heat than light-colored surfaces during the day, but radiate it more rapidly at night. Mulches reduce the soil heat flux either upward or downward, thereby moderating the diurnal soil temperature fluctuations, which are minimal below a depth of 20 cm. Soil temperature also fluctuates in a sinusoidal fashion from summer to winter. The temperature change is greatest at the surface and is attenuated with depth.

4

How the Soil Supplies Nutrients

Essential and Nonessential Elements for Grapevines

Most plants need 16 elements to grow normally and reproduce. Some of these elements are required in relatively large concentrations, ideally >1,000 mg/kg (0.1%) in the dry matter (DM); these are called *macronutrients* (table 4.1). The others, called *micronutrients* (table 4.1), generally are required in concentrations <100 mg/kg DM (0.01%).

Of the essential elements, C and O are supplied as CO_2 from the atmosphere, whereas H and O are supplied in H_2O from the atmosphere and water sources. Chlorine is also abundant in the air and oceans as the Cl^- ion. Winds whip sea spray containing Cl, Na, Mg, Ca, and S into aerosols to be deposited by rain on the land or as "dry deposition" on vegetation. Nitrogen as N_2 gas in the atmosphere enters soil–plant systems primarily by "biological fixation" (section 4.2.2.1), although small amounts are also deposited as NH_4^+ and NO_3^- ions from the air. Cobalt (Co) is essential for biological N_2 fixation in legumes and blue-green algae. For the remaining essential elements, the major source is minerals that weather in the soil and parent material.

Another term frequently used is *trace element*, which can include both essential and nonessential elements. A trace element normally occurs at a concentration <1,000 mg/kg in the soil. There are three categories of trace elements:

1. The essential *micronutrients* Cu, Zn, Mn, B, and Mo, which are beneficial at normal concentrations in the plant (ranging from 0.1 mg/kg for Mo to 100 mg/kg for Mn) but which become toxic at higher concentrations. Iron is the only micronutrient that is not strictly a trace element.
2. Elements such as chromium (Cr), selenium (Se), iodine (I), and Co that are not essential for plants, but are essential for animals.
3. Elements such as arsenic (As), mercury (Hg), cadium (Cd), lead (Pb), and nickel (Ni), which are not required by plants or animals and are toxic to either group at concentrations in the organism greater than a few mg/kg.

79

Table 4.1   *Macro- and Micronutrients, Their Chemical Symbols, and Common Ionic Forms in the Soil*

Macronutrient (>1,000 mg/kg)[a]	Common Ionic Forms in Soil	Micronutrient (<100 mg/kg)[a]	Common Ionic Forms in Soil
Carbon (C)	HCO_3^-, CO_3^{2-}	Iron (Fe)	Fe^{3+} (sometimes Fe^{2+})
Hydrogen (H)	H^+	Manganese (Mn)	Mn^{4+} (sometimes Mn^{2+})
Oxygen (O)	H_2O and many ions, e.g., OH^-, NO_3^-, SO_4^{2-}	Zinc (Zn)	Zn^{2+}
Nitrogen (N)	NH_4^+, NO_3^-	Copper (Cu)	Cu^{2+}
Phosphorus (P)	$H_2PO_4^-$, HPO_4^{2-}	Boron (B)	$B(OH)_4^-$
Sulfur (S)	SO_4^{2-}	Molybdenum (Mo)	MoO_4^{2-}
Calcium (Ca)	Ca^{2+}		
Magnesium (Mg)	Mg^{2+}		
Potassium (K)	K^+		
Chlorine (Cl)	Cl^-		

[a]Units of measurement are discussed in box 4.1.

Trace elements in the soil are normally derived from the parent material. Examples of concentrations of trace elements in soils derived from different parent materials are given in table 4.2. The rock *serpentinite* forms the parent material of soils at the northern end of the Napa Valley, California. Soils formed on serpentinite adversely affect the growth of grapevines because of their high Ni content and also because the macronutrient Mg is too high relative to Ca and K. Trace elements are also deposited naturally from the atmosphere, especially from volcanic eruptions and in dust and sea spray. In the twentieth century, however, anthropogenic sources of trace elements have increased markedly, especially from mining, industrial processing, motor vehicles, agricultural chemicals, and urban waste. These sources have increased trace element loadings in soil–plant systems, via the atmosphere and through direct solid and/or liquid applications (e.g., spread-

Table 4.2   *Trace Element Contents[a] of Soils Formed on Different Rock Types*

Element	Limestone and Dolomite	Serpentinite	Andesite	Granite	Sandstone
Basic	◄——————————————————————————►				Acidic
Co	0.2	160	16	<4	<4
Ni	40	1600	20	20	30
Cr	22	6000	120	10	60
Mo	0.8	2	<2	<2	<2
Cu	8	40	20	<20	<20
Mn	2200	6000	1600	1400	400
Pb	18	2	12	36	24

[a]In kg/ha
Source: Compiled from Mason (1966) and White (1997)

ing animal waste and biosolids). These inputs must be carefully monitored so that the potentially toxic elements do not exceed safe levels.

Concentrations of elements in soils and plants can be expressed in several different units of measurement. It is important to know the meaning of these units, and the relationships between laboratory and field measurements, as explained in box 4.1.

| Box 4.1 | *Ways of Expressing Concentrations of Elements* |

The concentration of an element (e.g., K) in soil or plant material is often expressed as μg K/g soil (or plant material). Note that

$$1 \ \mu\text{g K/g} = 1 \ \text{mg K/kg} = 1 \ \text{part per million (ppm K)} \qquad \text{(B4.1.1)}$$

From equation B3.4.4 (box 3.4), for a soil of average $BD = 1.33 \ \text{Mg/m}^3$, we know that the mass M of o.d. soil per ha-0.15 m is 1995 Mg. We can use this relationship to convert from laboratory results (mg/kg) to an element content on a field scale as follows:

$$1 \ \text{mg K/kg soil} = 1.995 \ \text{kg K/ha-0.15 m} \cong 2 \ \text{kg K/ha-0.15 m} \qquad \text{(B4.1.2)}$$

An exact conversion for any given soil can be made if the soil BD is measured. If the soil depth of interest is >0.15 m, the conversion factor in this equation is correspondingly greater. For example, suppose the average mineral N concentration in a vineyard soil was measured to a 1-m depth. The conversion equation would be

$$1 \ \text{mg N/kg soil} = \frac{2}{0.15} = 13.3 \ \text{kg N/ha-m} \qquad \text{(B4.1.3)}$$

The concentration of an element in the soil is sometimes expressed as a percentage. For example, 1% C may be written as

$$1\% \ \text{C} = \frac{1 \ \text{g C}}{100 \ \text{g soil}} = \frac{0.01 \ \text{kg C}}{\text{kg soil}} = 19,995 \ \text{kg C/ha-0.15 m} \qquad \text{(B4.1.4)}$$

Therefore,

$$1\% \ \text{C} \cong 20,000 \ \text{kg C/ha-0.15 m} \qquad \text{(B4.1.5)}$$

Fresh or air-dry soil samples are usually used for the analysis of "available" elements. Analysis of total element concentration may be done on o.d. soils. But all analytical values for soils and plants should be expressed on an o.d. basis. Plant material should be dried to constant weight at a temperature of 70°C. The gravimetric water content θ_g (see equation 3.3) is used to correct a soil analysis to an o.d. basis. For example, suppose the soil P content is measured as 0.1 P% (air-dry basis), and θ_g for the air-dry soil is 10%. Then

$$\% \ \text{P (o.d.soil)} = 0.1\% \ \text{P (air-dry soil)} \times 110/100 = 0.11\% \ \text{P (o.d. soil)} \qquad \text{(B4.1.6)}$$

Elemental analyses of organic manures and biosolids are often quoted as a percentage of fresh or "wet" weight. To convert to a DM basis, we need to measure the dry matter percentage of the fresh weight, and calculate the element concentration as follows:

$$\% \ \text{P (DM)} = \frac{\text{g P}}{100 \ \text{g fresh weight}} \times \frac{100 \ \text{g fresh weight}}{\text{g DM}} \qquad \text{(B4.1.7)}$$

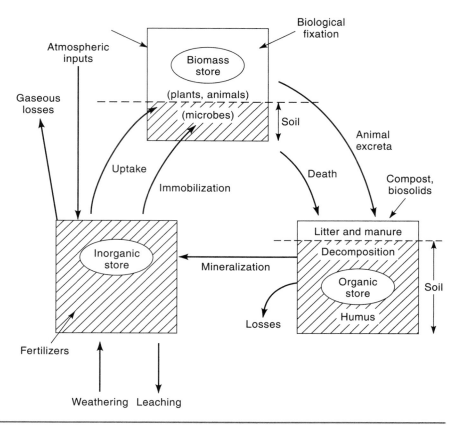

Figure 4.1   A generalized nutrient cycle.

Nutrient Cycling

Nutrients are continually cycling in the soil–plant system as plants and other organisms go through repeated cycles of growth, death, and decay. In this dynamic process, nutrients flow among these three major stores: an *inorganic store*, chiefly the soil, a *biomass store*, comprising organisms living above and below the ground, and an *organic store*, on and in the soil, formed by the dead residues and excreta of living organisms.

The general nutrient cycle is illustrated in figure 4.1. The relative size of the nutrient stores differs for each element, as does the importance of the pathways for inputs and outputs, and the partitioning of an element between "soil" and "nonsoil." Within each store, nutrients can undergo transformations, changing from one form (e.g., NH_4^+) to another (NO_3^-).

Differences also exist in the way the elements are distributed in the soil. Elements released at depth by mineral weathering are absorbed by plant roots and transported to the shoots, to be deposited finally on the soil surface in litter or excreta. Accessions from the atmosphere add to the surface store. N always accumulates in the organic-rich A horizon, and the total N content declines with depth, as shown in figure 4.2. Exchangeable NH_4^+ ions are retained in the A horizon, but NO_3^- ions can be leached and accumulate at depth. Although P distribution

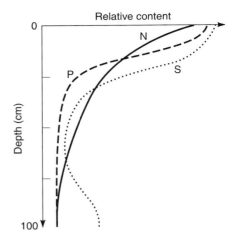

Figure 4.2   Changes in relative content of soil N, P, and S with depth.

is similar to N, the decline with depth is more abrupt because of the immobility of phosphate ions in the soil (exceptions can occur in very sandy soils, such as in the coastal regions of southwest Western Australia). Sulfur, like N, accumulates mainly in the A horizon, but SO_4^{2-} ions produced by mineralization of organic S may leach and accumulate in the subsoil (fig. 4.2). The distribution of most cations—Fe^{3+}, Ca^{2+}, Mg^{2+}, Mn^{2+}, Cu^{2+}, and Zn^{2+}, for example—usually correlates with clay accumulation in the soil profile. But cations such as Cu^{2+}, Zn^{2+}, Fe^{3+}, Al^{3+}, and Mn^{2+}, which have a high affinity for organic matter (section 2.3.4.2), also accumulate in the organic-rich parts of the soil profile.

4.2.1   *The Inorganic Store*

Plant roots normally access only a small fraction of the soil's inorganic store of the macronutrients P, Ca, Mg, and K and the micronutrients Fe, Mn, Zn, Cu, B, and Mo in a single growing season. This fraction is drawn from an *available pool*, made up of ions in the *soil solution* and *exchangeable ions* adsorbed by clay minerals and organic matter. The remainder of the inorganic store (the unavailable pool) exists as (1) sparingly soluble compounds like gypsum, calcite, and magnesite, (2) insoluble inorganic compounds of P, Fe, Zn, and Cu, (3) nonexchangeable ions in clay lattices (e.g., K^+), or (4) insoluble oxides (Fe, Mn, and Mo).

Inputs to the available pool occur by

1. weathering of soil and rock minerals (chapter 1)
2. rainfall and dry deposition (chapter 5)
3. mineralization of organic matter and excreta (chapter 2)
4. transfers from the unavailable pool (section 4.4.3)
5. application of manures and fertilizers (chapter 5).

4.2.2   *The Biomass Store*

The amount of nutrients stored in plants depends on the mass of vegetation produced. This varies greatly, for example, from a mature tropical forest to a cultivated crop, such as grapevines. The biomass C in a vineyard depends on many

Table 4.3   ***Biomass Nutrient Stores for Vines Compared to Other Vegetation Types***

Vegetation type	Nutrient Store (kg/ha)			
	N	K	Ca	Mg
Nonirrigated 18-yr-old vines (Cabernet Sauvignon), 1120 vines/ha, at harvest (including roots)	51	57		
Irrigated 10-yr-old vines (Chenin Blanc), 1680 vines/ha				
at bud burst (including roots)	215	124		
at harvest (including roots)	366	423		
Dry grassland (including roots and litter)	46	105	13	24
Mature rain forest (including roots and litter)	2040	910	2670	350

Source: After Williams and Biscay (1991), Mullins et al. (1992), and White (1997)

factors such as the variety, planting density, and vigor of the vines, as well as cultural practices (irrigation, pruning, leaf pulling, and fruit thinning). For example, the measured biomass C in a nonirrigated vineyard (1120 vines/ha) at fruit maturity was 5.2 t C/ha (all plant parts) (Williams and Biscay 1991). Of this biomass C, only about 15–20% would recycle annually. Examples of biomass stores for some of the macronutrients in vines and other vegetation types are given in table 4.3.

Cover crops in vineyards add considerably to the biomass store and are an important source of N if a legume is present. All biomass (above- and belowground) is important in nutrient cycling, irrespective of the size of the store, because the residues of living organisms and their excreta are the substrate on which the soil reducers and decomposers feed (section 2.3.2). It is difficult to estimate the size of the root biomass, because roots are intimately associated with the soil and can extend to a depth of several meters. Generally the root store amounts to 50–100% of the aboveground store.

Of the soil organisms, earthworms and the microbial biomass are the two most important nutrient stores. The microbial biomass has been identified as the "eye of the needle" through which all the C returned to the soil must eventually pass. Within this microbial biomass, the mean C:N ratio of bacteria is ca. 4 and that of fungi is ca. 10 (White 1997), so the biomass is also an important store of N (100–500 kg N/ha). Microbial biomass P values range from ca. 25 to 100 kg P/ha. A few species of microorganisms are exceptional in being able to reduce molecular N_2 to NH_3 and incorporate it into amino acids for protein synthesis through a process called *biological N_2 fixation.*

4.2.2.1   *Biological N_2 Fixation*

In agro-ecosystems, wherever legumes such as clover or lucerne are grown, the input of biologically fixed N is an important component of the N cycle. The N_2 fixed by legume cover crops can provide a significant input of N in vineyards. The microorganisms responsible for fixation live either independently in the soil as *free-living* organisms or in *symbiotic association* with plants.

Symbiotic N_2 fixation. Symbiosis denotes the cohabitation of two unrelated organisms that benefit mutually from the close association. The most important symbiotic N_2 fixers in vineyards are species of the family Fabaceae (Leguminosae) that form root nodules. The invading organism, or endophyte, is a bacterium of the genera *Rhizobium* and *Bradyrhizobium* that normally lives heterotrophically in the soil. When the root of a suitable host plant appears, the bacteria invade through the root hairs. The invaded cells and their neighbors in the host tissue respond by dividing and swelling to form a nodule. Within the nodule, the bacteria gain carbohydrates and energy from the host, and supply amino acids formed from the reduced N to the host. The biochemistry of this process is summarized in box 4.2.

Other symbiotic associations involve nonlegumes, such as the genera *Alnus* (alder), *Myrica* (bog myrtle), *Elaeagnus,* and *Casuarina.* Here the endophyte is usually an actinomycete (genus *Frankia*). Lichens are a symbiotic association between a fungus and a blue-green alga.

Because of the environmental and nutritional constraints shown in figure B4.2.1 and the inherent differences in the effectiveness of host–bacterial strain combinations, the quantity of N fixed is highly variable. For those legumes commonly used in vineyard cover crops, such as clovers (*Trifolium* spp.), medics (*Medicago* spp.), field beans (*Vicia* spp.), and field peas (*Pisum* spp.), the amount of N fixed annually is generally 40–60 kg N/ha. The legume N is made available when the nodules are sloughed off or when the legume residues decompose in the soil. Where animals graze, most of the N is returned through their excreta.

N_2 fixation by free-living organisms. These organisms are either photosynthetic or live heterotrophically. N_2 fixation in aerobic organisms requires the reduced form of the nitrogenase enzyme to be protected from O_2 (box 4.2), even though O_2 is required for the cell's respiration and production of energy-rich ATP. For this reason, the ability to fix N_2 is not widespread among nonphotosynthetic, aerobic organisms. At high concentrations of mineral N, the N_2 fixers are outcompeted by more aggressive heterotrophic bacteria in the soil. Thus, the rhizosphere of plant roots, with its abundance of exudates of high C:N ratio, is a favored habitat for free-living N_2 fixing organisms. Many of these organisms and associated nitrogenase activity have been identified in the rhizosphere of crops such as wheat, maize, pasture grasses, and sugar cane, but their significance for grapevines is unknown. This is called *associative* N_2 fixation.

Photosynthetic organisms make the greatest contribution to nonsymbiotic N_2 fixation; for example, up to 50 kg N/ha/yr is fixed by blue-green algae (Cyanobacteria) and photosynthetic bacteria in paddy rice fields. Estimates for other farming systems are in the range of 0–10 kg N/ha/yr.

4.2.3   *The Organic Store*

The nutrient cycle shown in figure 4.1 is completed by the decomposition of litter and animal excreta by soil fauna and microorganisms, as discussed in chapter 2. Ninety-eight percent or more of soil N is in organic molecules, and the metabolism of this N by soil microorganisms is intimately associated with that of C. Similarly, a substantial proportion of soil P and S exists in organic forms, which undergo mineralization and immobilization. Because of this accumulation in an organic form, replenished mainly by inputs at the soil surface, the distribution of N, P, and S generally follows the trends shown in figure 4.2.

N$_2$ reduction requires much energy. Approximately 16 moles of ATP are required per mole of N$_2$ reduced (ATP stands for the energy-carrying molecule adenosine triphosphate). The *nitrogenase* enzyme catalyzes the splitting of the dinitrogen molecule (N $\equiv$ N) and the reduction of each part to NH$_3$. NH$_3$ very rapidly combines with C compounds to form amino acids, the building blocks of proteins.

The amount of N fixed in symbiotic associations, especially those involving legumes, is much greater than that fixed by free-living organisms. The host plant supplies a large amount of energy through the oxidation of carbohydrates. The bacteria in the nodule (now called bacteroids) are supplied with O$_2$ by the pigment *leghemoglobin*, which has a very high affinity for O$_2$. By this means, the bacteroids can continue to respire, but the O$_2$ partial pressure at the surface of the bacteroids, where N$_2$ reduction occurs, is kept very low. When N$_2$ fixation is active in a nodule, the tissue has a characteristic pinkish-red color because of the functional leghemoglobin; this can be assessed by cutting open nodules detached from legumes in the field.

Successful N$_2$ fixation depends on factors that operate at any one of several points in the complex sequence of root hair infection, nodule initiation, and N$_2$ reduction. The main factors and how they operate are summarized in figure B4.2.1.

Figure B4.2.1 Factors affecting legume nodulation and the effectiveness of N$_2$ fixation.

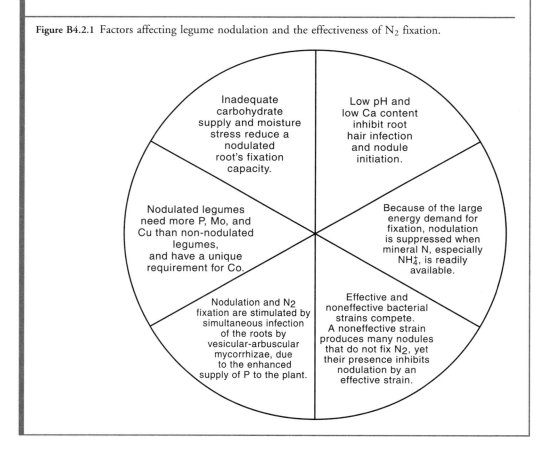

Important Transformations of N: Mineralization and Immobilization

Depending on the soil and environmental conditions, the quantity of N in the vine root zone (ca. 1 m deep) ranges from 1 to 10 t/ha, most of which occurs in the top 15–25 cm. Organic N (in proteins, nucleic acids, and complex ligno-protein compounds) is transformed to NH_4^+ ions during the decomposition of organic matter by heterotrophic microorganisms—the process of *ammonification.* A simplified description of the process is

$$\text{Organic N} \rightarrow NH_4^+ + OH^- \tag{4.1}$$

which indicates an alkalizing reaction. This reaction may take place under aerobic or anaerobic conditions, but it is slower in the latter case because less energy is available for microbial growth.

In well-aerated soils, ammonification is usually followed by the oxidation of NH_4^+ to NO_3^- by autotrophic bacteria. The principal nitrifying organisms are of the genera *Nitrosomonas* and *Nitrobacter*. These organisms derive energy for growth solely from the oxidation of NH_4^+, which is coupled to the reduction of CO_2 for the synthesis of complex C compounds. The oxidation occurs in two steps:

$$NH_4^+ + \frac{3}{2} O_2 \rightarrow NO_2^- + 2H^+ + H_2O + \text{Energy} \tag{4.2}$$

is carried out by *Nitrosomonas* species and

$$NO_2^- + \frac{1}{2} O_2 \leftrightarrow NO_3^- + \text{Energy} \tag{4.3}$$

is carried out by *Nitrobacter* species. The overall process, called *nitrification,* is represented by the reaction

$$NH_4^+ + 2O_2 \rightarrow NO_3^- + H_2O + 2H^+ + \text{Energy} \tag{4.4}$$

NH_4^+ and NO_3^- ions comprise the pool of *mineral N* on which plants feed. The incorporation of mineral N into complex N compounds in living organisms is called *immobilization* (section 2.3.1.1). Thus, both plants and nitrifying organisms compete with heterotrophic soil organisms for the limited pool of mineral N in soil. The balance between microbial mineralization and immobilization is determined primarily by the C:N ratio of the substrate (section 2.3.1.2). The critical C:N ratio below which net mineralization occurs is 20–25. This figure is used to determine the supply of soil N to grapevines by mineralization (chapter 5).

In addition to the NH_4^+ supply, soil temperature, moisture, pH, and O_2 status influence the growth and activity of nitrifying organisms in the following ways:

- The optimum temperature is between 30 and 35°C, and nitrification is very slow at temperatures <5°C.
- The optimum soil water content is about 60% of field capacity. Prolonged desiccation kills the bacteria, but sufficient bacteria survive short periods for the rate of nitrification to increase rapidly during the flush of decomposition that follows the rewetting of an air-dry soil.
- *Nitrosomonas* is more sensitive to low pH than *Nitrobacter*. But measurements of the short-term nitrification rate at nonlimiting NH_4^+ concentrations indicate that soil nitrifiers adapt to the prevailing pH, so

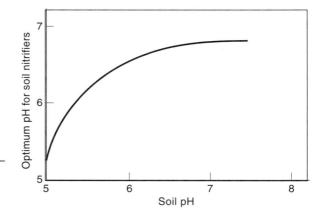

Figure 4.3   Optimum pH for soil nitrifiers (redrawn from White 1997).

that the optimum pH for nitrification is close to the soil pH (fig. 4.3). Thus, the optimum pH in any soil is unlikely to exceed 6.6–6.8.

- Autotrophic nitrifiers are strict aerobes. Once the O_2 partial pressure in the soil air falls below 0.4 kPa, nitrification ceases. Under these conditions, reduction of NO_3^- or *denitrification* may occur (section 5.6.1).

Ammonification, nitrification, and immobilization are key processes in the overall cycling of N among the atmosphere, soil, and plants, called the *nitrogen cycle* (fig. 4.4).

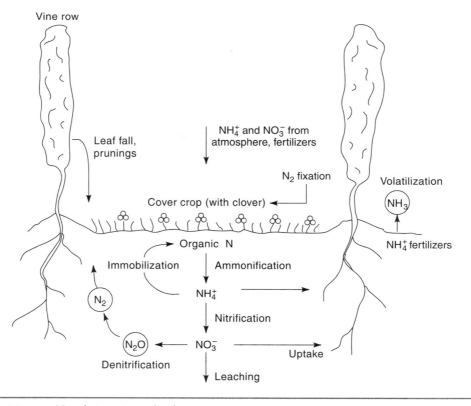

Figure 4.4   N cycle in a vineyard soil system.

4.4   *Mineralization and Immobilization of P and S*

4.4.1   *Phosphorus*

Soil P is normally the major reservoir of P in the P cycle between the soil and plants. The soil P content ranges from 500 to 2,500 kg P/ha, of which 30–85% may exist as organic P with the remainder in inorganic forms. Very little of this P is immediately available to plants. As in the case of N, mineralization of organic P occurs when microorganisms oxidize C substrates to obtain energy for growth. Some P-rich substrates, such as nucleic acids and nucleotides, are rapidly mineralized to orthophosphate. But when the C:P ratio of the substrate exceeds 60–100, net immobilization occurs because there is insufficient P to satisfy the microorganisms, especially bacteria, which have a high P requirement per unit weight. Thus, in ecosystems with only small P inputs, such as natural grasslands, the soil organisms are highly competitive with higher plants for P. Bacterial residues rich in P comprise mainly insoluble Ca, Fe, and Al salts of inositol hexaphosphate, called *phytates*. These are strongly adsorbed to soil particles and hence are protected from further decomposition.

The main form of P in the soil solution that is available to plants is orthophosphate: the ions $H_2PO_4^-$ and HPO_4^{2-}, whose proportions are controlled by the reversible equilibrium

$$H_2PO_4^- \longleftrightarrow HPO_4^{2-} + H^+ \tag{4.5}$$

At pH 7.2, the proportions of the monovalent and divalent orthophosphate ions are equal. If the pH falls to 5, the ratio of $H_2PO_4^-$ to HPO_4^{2-} is 100:1, whereas at pH 9 the ratio of $H_2PO_4^-$ to HPO_4^{2-} is 1:100. Mycorrhizae can also influence plant P uptake (section 4.7.3.2).

4.4.2   *Sulfur*

Soil S is derived originally from sulfide minerals in rocks that are oxidized to SO_4-S on weathering. This sulfate is absorbed by plants and returned to the soil in organic residues. Normally, therefore, soil S occurs mainly in an organic form amounting to 200–2,000 kg S/ha. Organic S is mineralized to SO_4^{2-} ions during microbial decomposition. The C:S ratio of well-humified organic matter is 50–150, which sets the approximate critical value for net mineralization of S.

Many vineyard soils are clean-cultivated. This inevitably results in lower organic matter contents than these soils would have in their natural state, so the amount of S (as well as N and P) released by organic matter decomposition is small. However, elemental S is regularly applied as a fungicidal "wettable powder," and any S residue that reaches the soil will be oxidized to SO_4^{2-} by *Thiobacillus* bacteria (section 5.4.2.2). Although SO_4-S is adsorbed by sesquioxides to some extent, much will leach and tend to accumulate in the subsoil, as, for example, in the Lateritic Podzolic Soils under vines in the Margaret River region of Western Australia.

4.4.3   *Labile and Nonlabile Pools*

$H_2PO_4^-$, HPO_4^{2-}, and SO_4^{2-} ions are adsorbed by mineral particles from the soil solution. Phosphate ions are much more strongly adsorbed than sulfate, for

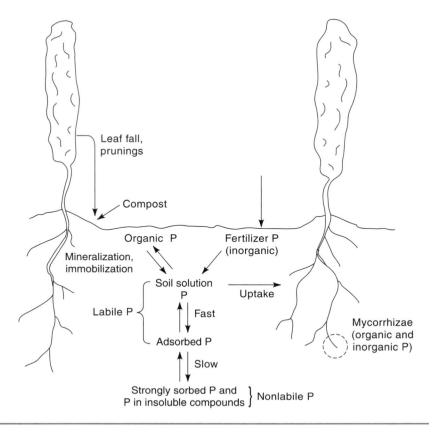

Figure 4.5   P cycle for a vineyard soil system.

the reasons discussed in section 4.5.4. This means that the P concentration in the soil solution is normally very low (<0.01 mg/L). Sorbed ions that are easily desorbed, plus those ions already in solution, collectively comprise the *labile pool* of a nutrient. Those ions that are very strongly adsorbed or trapped in insoluble compounds and recalcitrant organic forms comprise the *nonlabile pool*. Nutrients in the labile pool are considered to be "plant-available."

In addition to the adsorbed forms, P may exist in insoluble compounds with Fe, Al, or Ca, especially around dissolving fertilizer granules where high P concentrations and low pH exist for a time (section 5.4.2.1). In soils of semiarid to arid regions, sulfate can accumulate in the profile as gypsum ($CaSO_4.2H_2O$). Phosphorus cycling in the soil–plant system and transformations between labile and nonlabile forms of P in the soil are illustrated in figure 4.5.

4.5   *Partitioning of Ions Between the Solid and Solution Phases*

4.5.1   *The Diffuse Double Layer*

The way in which nutrient ions are retained, or released, and how the ionic composition can affect a soil's physical properties depend very much on the behavior of cations and anions at organic and mineral surfaces. A basic understanding of surface phenomena should, therefore, enable a viticulturalist to better manage vineyard soils.

As indicated in chapter 2, the crystalline clay minerals and sesquioxides have electric charges that attract cations and anions from the surrounding soil solution. Soil organic matter (*SOM*) also has a negative charge that varies with soil pH. The permanent negative charge in the clay minerals, which have thin laminar structures (see box 2.4), acts as if it were spread over the planar surfaces of the crystals. The hydrophilic carboxylic and phenolic groups in organic matter also point into the soil solution. This negative charge attracts a surplus of cations (the *counterions*) into the solution adjacent to the surface. At the same time, anions (the *co-ions*) are repelled to create a deficit of these ions in solution near the surface. The combination of fixed surface charges and mobile ions of opposite charge in solution comprises a *double layer*. Overall, the double layer is electrically neutral.

However, the electrostatic force between the fixed charges and mobile ions in the double layer is diminished by the high dielectric constant of water. The attractive (or repulsive) force acting on the mobile ions is also opposed by a diffusive force pulling these ions away from a region of high concentration to one of lower concentration. At equilibrium, the electrostatic force per unit area of surface is just balanced by the difference in osmotic force per unit area between the surface and the soil solution well removed from the surface. The result is a nonuniform distribution of cations and anions in solution, illustrated in figure 4.6a, in which the ion concentrations change exponentially as the surface is approached (fig. 4.6b). The combination of fixed surface charges and distributed mobile charges is called a *diffuse double layer* (*DDL*). Such a *DDL* develops at surfaces where the charge is permanent (constant), such as clay mineral planar surfaces, or the charge is variable, depending on the solution pH, such as clay mineral edge faces, oxide surfaces, and organic matter. Box 4.3 gives an example of how the charge on a kaolinite edge face, or an oxide surface, changes with pH.

4.5.2   *How the Diffuse Double Layer Can Change*

The *DDL* at charged surfaces is dynamic. Its effective thickness is measured by the average distance from the surface over which the concentration of the mobile

Figure 4.6   (a) The distribution of mobile cations (+) and anions (-) with distance from a negatively charged surface. (b) Ion concentrations relative to the bulk solution concentration C_o (White 1997). Reproduced with permission of Blackwell Science Ltd.

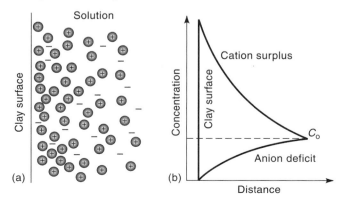

Box 4.3   *pH-Dependent Charge on Mineral Surfaces*

Oxygen and OH groups coordinated to Si and Al atoms in a kaolinite crystal are exposed at the crystal edges, and so have potentially unneutralized charges. These are satisfied by H^+ ions from solution, which become tightly bound to the O and OH groups. The H^+ ions enter into the surface plane of atoms and change the charge on that surface, as shown in figure B4.3.1. The surface groups behave like weakly dissociating acids (see appendix 3), and a reversible equilibrium involving H^+ ions is set up between the surface and solution. At low pH (high H^+ ion activity), the surface acquires a net positive charge. As the pH rises (lower H^+ ion activity), the tendency for H^+ ions to dissociate from the surface increases, so that at high pH the surface charge becomes negative. At some intermediate pH, the surface bears no net charge. This pH defines the *point of zero charge* (*PZC*).

Figure B4.3.1 pH-dependent charges at a kaolinite edge face (White 1997). Reproduced with permission of Blackwell Science Ltd.

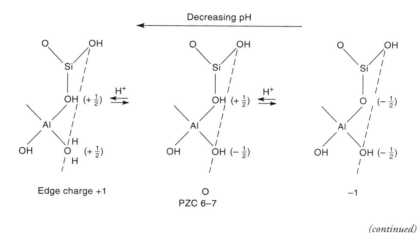

(continued)

---

ions is affected by the fixed surface charges. The electrostatic force that attracts mobile cations to a negatively charged surface, or repels anions, is directly proportional to the charge on the ion. For cations, the force decreases in the order $Al^{3+} > Ca^{2+} \cong Mg^{2+} > Na^+ \cong K^+$. Thus, for a fixed concentration of salts in the soil solution, we would expect a *DDL* made up of Al^{3+} ions to be much thinner, or more compressed, than that made up of Na^+ ions.

In addition to the force pulling cations to the surface, we must consider the countervailing osmotic force pulling them out into the bulk solution. Diffusion outward depends on the difference in concentration for *individual* ion species between the surface region and the bulk solution. The concentration of whichever cation is attracted to the clay surface (e.g., Na^+) is determined by the charge on the clay. Therefore, the difference in concentration across the *DDL*, and hence the outward diffusive force, will change as the bulk solution concentration changes.

Box 4.3 *(continued)*

The same processes of association and dissociation of H^+ ions, dependent on the solution pH, occur at the surfaces of sesquioxides. The acid strength of the surface groups changes according to the mineral's composition, so that different minerals have different *PZC*s. For example, the *PZC* for quartz is 2–3, kaolinite edge faces is 6–7, goethite (FeOOH) is 7–8, and gibbsite (Al(OH)$_3$) is 8–9. This means that goethite is positively charged up to ca. pH 7, and gibbsite positively charged up to ca. pH 8. However, the *PZC* values for these minerals can change if cations like Ca^{2+} or some of the trace elements, or anions like $H_2PO_4^-$, form complexes with the surface (section 4.5.4). In soil, these variably charged surfaces coexist with the constant charge surfaces of clay minerals. Indeed, positively charged oxide coatings commonly form on negatively charged clay mineral surfaces and provide sites for anion adsorption (fig. B4.3.2).

Figure B4.3.2 Sites for anion adsorption on negatively charged clay crystals (White 1997). Reproduced with permission of Blackwell Science Ltd.

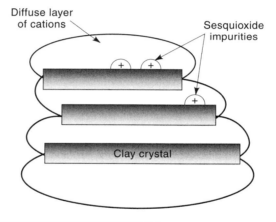

The higher the bulk solution concentration, the less the outward force, hence the more compressed the *DDL*.

The resultant of these opposing forces, which depend on ion charges and concentrations, determines the effective thickness of the *DDL*. For solutions of mixed salts such as a soil solution, we use a property called the *ionic strength* (I) as a generic measure of ion concentration, modified by ion charge. The thickness d_{eff} of the *DDL* can then be calculated from the approximate relationship

$$d_{eff} \propto \frac{1}{\sqrt{I}} \tag{4.6}$$

which, from the definition of ionic strength given in box 4.4, expands to

$$d_{eff} \propto \frac{1}{z\sqrt{C}} \tag{4.7}$$

Box 4.4   *Ion Activity and Ionic Strength*

Concentration refers to the number of molecules or ions per unit volume. An individual ion experiences weak forces because of its interaction with water molecules (the formation of a hydration shell) and with other ions of the same or opposite charge. Effectively, the ability of the ion to engage in chemical reactions is decreased, relative to what is expected when it is present at a particular concentration. This effect is accounted for by defining the *activity a* of the ion, which is related to its concentration C by the equation

$$a = fC \qquad\qquad\qquad (B4.4.1)$$

where f, with values from 0 to 1, is the activity coefficient of that ion species. In very dilute solutions where the interaction effects are small, f approaches 1, and a is approximately equal to C. An extensive theory about how to calculate activity coefficients exists. A key variable in determining the value of f for an ion in solution is I, which incorporates the cumulative effect on f of the charge and concentration of all the ion species (1 to n) in solution, through the equation

$$I = 0.5 \sum_{i=1}^{i=n} C_i z_i^2 \qquad\qquad (B4.4.2)$$

Equation 4.7 shows that the effective thickness of the *DDL* is inversely proportional to the valency z of the counterion and inversely proportional to the square root of the bulk solution concentration C. Some examples of d_{eff} values are given in table 4.4.

Changes in *DDL* thickness are discussed here in terms of the forces acting on solute ions. But we should remember that these ions are immersed in a "sea" of water molecules. A change in the type of ion and its concentration in the *DDL* results also in a change in the activity of water molecules. Water molecules diffuse into, or out of, the *DDL* in response to their own activity gradient across the *DDL*. For example, if the ion concentration in the bulk solution decreases, the tendency for water molecules to diffuse into the *DDL* is increased. This leads to a *swelling pressure*, which depends on the type of cations in the *DDL* and the soil

---

**Table 4.4   *Effective Thickness of the* DDL *at a Clay Surface
in Salt Solutions of Different Concentrations***

Salt Solution of Concentration C (mol/L)	Effective Thickness d_{eff} of the *DDL* (nm)[a]	
	NaCl	CaCl$_2$
0.1	1.94	1.0
0.01	6.2	3.2
0.001	19.4	10.1

[a]1 nm = 10^{-9} m

Figure 1.2   Soil profile showing A,
              B, and C horizons.
              Note the color change
              among the sandy loam
              A horizon, the clay-
              enriched B horizon with
              Fe_2O_3 accumulation,
              and the pale-colored C
              horizon of weathering
              siltstone. Photograph by
              the author.

Figure 1.3   Granite boulders
              exposed by weathering
              in northeast Victoria.
              Photograph by the
              author.

Figure 1.4   A deep, coarse-textured, well-drained soil in the Calquenas region, south-central Chile. Photograph by the author.

Figure 1.5   A Rendzina soil profile in McLaren Vale, South Australia. Photograph by the author.

Figure 1.6   Old vines growing in weathered
              schist in Languedoc-Roussillon,
              southern France. Photograph by
              the author.

Figure 1.7   River gravel deposits used for
              viticulture in the Lodi District,
              Central Valley, California.
              Photograph by the author.

Figure 1.8   Glaciofluvial deposit on older weathering granite in the Calquenas region, south-central Chile. Photograph by the author.

Figure 1.10   Vines growing on limestone fragments near St Jean de Minervois, Languedoc-Roussillon. Photograph by the author.

Figure 2.12 An example of fungal mycelium and fruiting bodies. Photograph by the author.

Figure 2.14 Earthworm casts on the soil surface. Photograph by the author.

Figure 3.1 Friable tilth created by frost in cultivated soil. Photograph by the author.

(a)

(b)

(c)

(d)

(e)

Figure 3.2   (a) Surface soil crumb structure, class 0.5–1 cm. (b) Subangular blocky aggregates, class 2–3 cm. (c) Platy aggregates in the A2 horizon of a sodic soil, class 0.5–4 cm. (d) Prismatic aggregates in a clay subsoil, scale is 10 cm. (e) Columnar aggregates in a sodic subsoil, 12–15 cm across the columns. Photographs by the author.

Figure 3.5   An ironstone gravel layer at the top of the B horizon in a duplex soil on sandstone in the Yarra Valley, Australia. Photograph by the author.

Figure B4.6.1   A portable pH meter with electrodes in a soil suspension. Photograph by the author.

solution concentration. As discussed in chapter 3, the expansion or contraction of *DDLs* at clay surfaces is crucial in determining whether clay crystals remain flocculated. Further, the state of clay flocculation strongly influences the stability of soil aggregates and the soil's structure in general, a topic taken up again in chapter 7.

4.5.3   *Some Additional Effects on the DDL*

All ions in solution are hydrated. The radius of a hydrated ion is larger than that of an unhydrated ion. We saw in box 2.6 that, under the polarizing effect of fixed charges in a mineral surface, cations could either completely dehydrate to form tightly bound complexes with the surface or partially dehydrate to form less tightly bound complexes. In both cases, the cation is pulled close to the surface, effectively moving from the diffuse part of the *DDL* into a layer called the Stern layer immediately adjacent to the surface (fig. 4.7). This process is called *specific adsorption*. Specific adsorption forces add to the *nonspecific* electrostatic forces described previously. Overall, specific forces acting on the cations cause the *DDL* to become more compressed. For cations of the same valency, such as the series Li^+, Na^+, K^+, Rb^+, and Cs^+, the ease of dehydration of the cation increases as the atomic radius increases, from Li^+ through Cs^+. Consequently, the specific adsorption effect increases with the size of the unhydrated cation, and we find that Cs^+, for example, is much more strongly adsorbed than Na^+. The *DDL* of a Cs-saturated clay is more compressed than that of a Na-saturated clay, and hence Cs-clays flocculate much more readily than Na-clays.

4.5.4   *Anions and Charged Surfaces*

Anions such as Cl^-, SO_4^{2-} and HCO_3^- are normally repelled from negatively charged surfaces. However, if an anion has a strong chemical affinity for metal

---

Figure 4.7   Diagram of a diffuse double layer at a negatively charged clay surface showing Stern layer cations (after White 1997).

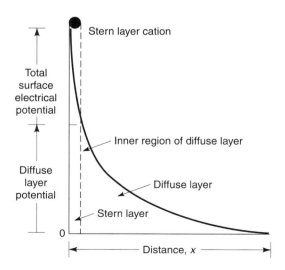

ions in the surface, such as $H_2PO_4^-$ has for Al^{3+} and Fe^{3+}, it may be specifically adsorbed at sites within the *DDL*. Such a possibility is shown in figure B4.3.2. The $H_2PO_4^-$ ion displaces one of the ligands (—OH) from the coordination shell of a metal ion (M) at the surface by a process of *ligand exchange*, described by the reaction

$$= M—OH + H_2PO_4^- \leftrightarrow = M—O_2(OH)_2P + OH^- \qquad (4.8)$$

This reaction may occur at the edge faces of kaolinite crystals, where Al is the metal cation, or in sesquioxide surfaces where Al and Fe are the metal cations. As demonstrated in box 4.3, such surfaces may have a net positive, neutral, or negative charge, depending on the solution pH. The $H_2PO_4^-$ ion can react also with a protonated site (—OH combined with H^+), as, for example, in the reaction

$$= M—OH_2^+ + H_2PO_4^- \leftrightarrow = M—O_2(OH)_2P + H_2O \qquad (4.9)$$

In both cases, the $H_2PO_4^-$ ion is very strongly adsorbed and is desorbed only in exchange for other anions with a strong affinity for the surface, such as OH^-, molybdate (MoO_4^{2-}), or silicate ($H_3SiO_4^-$), when these are present at a high enough concentration. Anions such as Cl^-, SO_4^{2-} and NO_3^- are *not* effective competitors with $H_2PO_4^-$ at these specific adsorption sites. However, they are attracted to positively charged sites by electrostatic forces—a process of nonspecific adsorption—and are exchanged for other nonspecifically adsorbed anions according to their relative concentrations in solution. An example is given by the following reaction:

$$= M—OH_2^+ Cl^- + NO_3^- \leftrightarrow = M—OH_2^+ NO_3^- + Cl^- \qquad (4.10)$$

Thus, anions can be specifically or nonspecifically adsorbed at the surfaces of sesquioxides, which may occur as discrete particles or as coatings on soil particles (Fig. B4.3.2). These anions can be exchanged for other anions in solution, according to reactions 4.8–4.10, which gives rise to an overall *anion exchange capacity* (*AEC*) for the surface. The significance of these reactions for the nutrition of grapevines is discussed in chapter 5.

4.6   *More Ion Exchange Reactions*

4.6.1   *Exchangeable Cations and Anions*

Negatively charged surfaces predominate in most viticultural soils. Cations held in the *DDL* (apart from those forming tight complexes) are called *exchangeable* because they can exchange rapidly with cations in the bulk solution. In all but the most acid or alkaline of soils, the major cations are Ca^{2+}, Mg^{2+}, K^+, and Na^+, usually with Ca^{2+} plus Mg^{2+} being more abundant than Na^+ plus K^+. In some viticultural areas, such as the northern end of the Napa Valley in California and the Marlborough district in New Zealand, Mg^{2+} may exceed Ca^{2+}. In some Australian regions, such as the Clare Valley and the Riverland of South Australia, exchangeable Na^+ plus Mg^{2+} can be higher than Ca^{2+}, which can predispose the soil to structural problems (chapter 7).

The amount of exchangeable NH_4^+ may vary, depending on the soil's nitrification activity. Relatively small amounts of other cations, such as the micronutrients Fe^{3+}, Cu^{2+}, Mn^{2+}, and Zn^{2+} and the "heavy metals" Cd^{2+}, Ni^{2+}, and Pb^{2+}, are also adsorbed. But chelation with organic compounds and precipitation in insoluble salts play a more prominent part in the retention of the last two groups of cations. Ions such as Cl^-, SO_4^{2-}, and NO_3^- are held as exchangeable anions in soils that are highly weathered and contain appreciable sesquioxides (positively charged). This is more common in the subsoil where the influence of organic matter is negligible.

The capacity of a surface to hold exchangeable cations depends on its charge per unit area. The most convenient surrogate for this variable is the moles of charge per unit mass (see box 2.5), which is a measure of the *cation exchange capacity* (*CEC*). Consider the reversible exchange reaction

$$(NH_4^+) + (Ca\text{-}clay) \leftrightarrow (NH_4\text{-}clay) + {}^1/_2(Ca^{2+}) \tag{4.11}$$

In this reaction, one mole of NH_4^+ charge in solution exchanges for one mole of Ca^{2+} charge on the clay. The terms (NH_4^+) and (Ca^{2+}) represent the molar activities of the ions in solution, and $(Ca\text{-}clay)$ and $(NH_4\text{-}clay)$ represent the activities of the adsorbed ions. In most cases, we may substitute concentrations for activities (box 4.4). According to the Law of Mass Action, the equilibrium will be displaced to the right (favoring displacement of Ca^{2+} ions into the solution) by increasing the concentration of free NH_4^+ ions or by removing Ca^{2+} ions from the site of the reaction. This principle is the basis of *CEC* measurements, as outlined in box 4.5.

Reversible reactions of the form of equation 4.11 can be written for several cation exchange equilibria in soil, for example

$$(Na^+) + (Ca\text{-}clay) \leftrightarrow (Na\text{-}clay) + {}^1/_2(Ca^{2+}) \tag{4.12}$$

$$(Na^+) + (Mg\text{-}clay) \leftrightarrow (Na\text{-}clay) + {}^1/_2(Mg^{2+}) \tag{4.13}$$

These reactions are discussed in chapter 7 in the context of soil sodicity.

Box 4.5	***Methods of Measuring the CEC of Clay Minerals and Soils***

Methods of measuring *CEC* fall into two groups (Rayment and Higginson 1992):

1. Exchangeable cations are displaced at a fixed pH by a concentrated salt solution, such as M NH_4Cl or M NH_4OOCH_3 at pH 7. Displacement is achieved by the high concentration of the "index" cation (NH_4^+). The index cation is then displaced by another cation (e.g., Na^+) and the amount of NH_4^+ adsorbed (cmol (+)/kg) is measured.
2. Exchangeable cations are displaced by an unbuffered salt solution at no fixed pH, and the amount of index cation that has been adsorbed (e.g., Ba^{2+} from 0.1M $BaCl_2$/0.1M NH_4Cl solution) is measured. Displacement is achieved because of the high affinity of Ba^{2+} for the surface.

If the concentration of exchangeable cations is to be measured in the extracts, a prewash with aqueous ethanol is used to remove any soluble salts. *CEC* values determined by methods of the first group are widely reported in soil survey data, and are used in some soil classifications. However, in soils with variably charged

(continued)

Box 4.5 *(continued)*

surfaces, the *CEC* at pH 7 often overestimates the *effective CEC* (*ECEC*), that is, the soil's *CEC* at its natural pH. For such soils, methods of the second group are preferred. Because different methods can give different values for the *CEC*, it is important to specify the method of measurement.

Typical values for the *CEC* of the common clay minerals are given in table B4.5.1. *CEC* values of viticultural soils containing these minerals will vary according to the proportion of each mineral in the clay fraction and the total clay content. For example, a Krasnozem with 50% clay comprising equal proportions of kaolinite (*CEC* = 10 cmol (+)/kg clay) and illite (*CEC* = 40 cmol (+)/kg clay) would have a *CEC* of 12.5 cmol (+)/kg soil. Organic matter in the A horizon would add to the *CEC* value.

Table B4.5.1 CEC *Values for the Common Groups of Clay Minerals*

Clay Mineral Group	*CEC* (cmol Charge (+)/kg)[a]
Kaolinite	3–20
Illite	10–40
Smectite (montmorillonite)	80–120
Vermiculite	100–150

[a]Although the charge on the clay is negative, it is measured as the number of moles of cation charge (+) adsorbed.

4.6.2   *Nonexchangeable Cations and Anions*

Exchangeable cations and anions are readily available to plants. When plants absorb these ions from the soil solution, exchangeable ions are desorbed to "buffer" the solution concentration against change. The desorbed ions are replaced on the surface by ions that are readily available in the solution, such as H^+ (for cations) or HCO_3^- (for anions). However, exchangeable ions of the macronutrients, such as Ca, Mg, K, P, and S, make up only a fraction of the total inorganic content of that nutrient in the soil. A proportion of the element may also exist in minerals, such as the feldspars, or in insoluble compounds such as $CaCO_3$, FeS_2, and $Ca_{10}(PO_4)_6(OH)_2$. For cations such as K^+, an important fraction is held in the interlayer spaces of hydrous mica clay minerals, as explained in section 2.2.4.2. In the interlayer spaces, the unhydrated K^+ ions, as well as NH_4^+ ions because of their similar size, are effectively "nonexchangeable" to cations in solution. But as the minerals weather, they begin to exfoliate, and the interlayer K^+ ions at the edges become more accessible for exchange (fig. 4.8). These K^+ ions, occupying "wedge-shaped" sites, are still held more strongly than the hydrated, exchangeable ions on the flat surfaces, but they are released as the concentration of exchangeable K^+ is depleted by plant uptake.

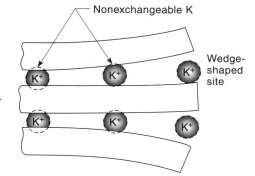

Figure 4.8 Exfoliation of a micaeous clay on weathering to form "wedge-shaped" sites that hold K+ ions (White 1997). Reproduced with permission of Blackwell Science Ltd.

The micronutrient cations Cu^{2+}, Fe^{3+}, Mn^{2+}, Zn^{2+}, and Co^{2+} are also classed as "nonexchangeable" when they form chelates with organic compounds (section 2.3.4.2). Similarly, when the nutrient anions of P, Mo, and B are specifically adsorbed on clays or oxides, they are not readily exchanged by anions such as Cl^- and NO_3^-. But as discussed in section 4.5.4, they are exchanged by other specifically adsorbed anions supplied at a high concentration in solution, or when the pH is raised. These forms of cations and anions comprise a nutrient pool of intermediate availability to plants—intermediate between the readily exchangeable forms and the insoluble precipitates, or unweathered rock minerals.

4.6.3 *Exchangeable Al^{3+} and Soil Acidity*

4.6.3.1 *Acid Clays*

A soil becomes acidic as Ca^{2+}, Mg^{2+}, K^+, and Na^+ ions are leached faster than they can be replaced by mineral weathering and atmospheric inputs. Initially, the exchangeable cations are replaced by H^+ ions, which are continually generated from carbonic acid (H_2CO_3) formed by CO_2 dissolving in the soil water:

$$CO_2 + H_2O \leftrightarrow H_2CO_3 \leftrightarrow H^+ + HCO_3^- \leftrightarrow H^+ + CO_3^{2-} \qquad (4.14)$$

Pure rainwater in equilibrium with CO_2 in the atmosphere has a pH of 5.65. But as respiration increases the partial pressure of CO_2 in the soil air, the reactions are driven to the right, producing more H^+ ions. The intensity of acidity that develops is measured by the soil pH, as outlined in box 4.6.

In soil, the small mobile H^+ ions invade clay mineral lattices. At a pH <4, mineral weathering is accelerated, releasing Al, SiO_2, and smaller amounts of Mg, K, Fe, and Mn. The SiO_2 combines with water to form weak silicic acid ($Si(OH)_4$), which leaches away, and the Al, Mg, K, and some Mn are retained, initially as exchangeable cations. With the exception of soils containing illitic clay, K^+ is lost by leaching, and the clay that remains is dominated by Al^{3+} with some Mg^{2+} (fig. 4.9). Hence an *acid clay* becomes an Al-clay. The hydrated Al^{3+} ions are exchangeable, and the amount present is usually measured by displacement in M KCl solution. Some H^+ ions, produced by the hydrolysis of the Al^{3+} ions (see next section), are also displaced. The sum of exchangeable Al^{3+} and H^+ defines the soil's *exchangeable acidity* (in cmol H^+/kg).

Box 4.6   *pH and Its Measurement in Soil*

pH is defined as the negative logarithm of H^+ ion activity (measured in solution as moles/L). For practical purposes, we equate activity to concentration in dilute solutions. The pH scale ranges from 0 to 14, with pH 7 indicating the neutral point (equal concentrations of H^+ and OH^- ions).

Soil pH can be measured in the field with a universal indicator and a color chart to an accuracy of $\pm$ 0.5 pH units. In the laboratory, soil pH is measured to an accuracy of $\pm$ 0.05 units in a soil suspension, using a glass electrode and a reference calomel electrode (fig. B4.6.1). The suspension is made by shaking one part by weight of soil with five parts by volume of distilled water. Robust glass-calomel electrodes have also been developed for direct use on moist soils in the field. Dilution of the soil solution with distilled water changes the soil pH. For soils containing mainly negatively charged clays and organic matter, the measured pH in water, which is the pH of the bulk solution, is higher than that of the undisturbed soil.

However, if soil is shaken with a solution of 0.01M $CaCl_2$, the measured soil pH changes little with dilution, because 0.01M $CaCl_2$ has an ionic strength (see box 4.4) similar to that of the average soil solution. Thus, many laboratories have now adopted the preferred method of measuring pH in 0.01M $CaCl_2$ at a soil:liquid ratio of 1:5. Because pH ($CaCl_2$) is generally 0.6–0.8 units lower than pH (H_2O) at the same soil:liquid ratio, it is important to know, when pH values are quoted, which method of measurement has been used.

Figure B4.6.1 A portable pH meter with electrodes in a soil suspension. Photograph by the author. See color insert.

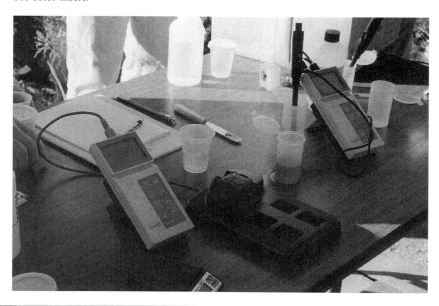

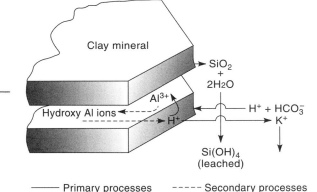

Figure 4.9   Weathering of a clay mineral under acid conditions to form an Al-clay (White 1997). Reproduced with permission of Blackwell Science Ltd.

——— Primary processes   ----- Secondary processes

High concentrations of exchangeable Al^{3+} have an adverse effect on the growth of grapevine roots (section 5.5.3).

4.6.3.2   *Base Saturation*

In calcareous soils, the sum of the exchangeable cations, expressed as Σ (Ca, Mg, K, Na) (meaning "sum of"), is invariably equal to the effective *CEC* because any deficit of exchangeable cations is made up by Ca^{2+} ions dissolved from $CaCO_3$. In noncalcareous soils, Σ (Ca, Mg, K, Na) is usually less than the effective *CEC*, the difference being made up by the exchangeable acidity, that is,

$$\text{Effective CEC} - \sum (\text{Ca, Mg, K, Na}) = (Al^{3+} + H^+) \qquad (4.15)$$

The ratio

$$\frac{\sum (\text{Ca, Mg, K, Na})}{\text{CEC}} \times 100$$

is commonly called the *percent base saturation*. Historically these cations have been called "basic" because they are the cations of strong bases in water, such as $NaOH$, KOH, $Mg(OH)_2$, and $Ca(OH)_2$. The term is useful to distinguish these cations from the "acidic" cations, such as Fe^{3+} and especially Al^{3+}, which hydrolyze in water to release H^+ ions in the pH range 4–6. For example, Al^{3+} in water is really a hexahydrated ion that hydrolyzes as follows:

$$Al(H_2O)_6{}^{3+} + H_2O \leftrightarrow [Al(OH)(H_2O)_5]^{2+} + H_3O^+ \qquad (4.16)$$

The ion on the right-hand side is called a hydroxy-aluminum ion. There are equal concentrations of the divalent and trivalent Al ions at pH 5, which is the pK value for this reversible reaction (see appendix 3).

4.6.3.3   *Titratable Acidity and pH Buffering Capacity*

As acid soils weather over many years, they produce appreciable amounts of exchangeable Al^{3+}, much of which hydrolyzes according to reaction 4.16. The hydroxy-Al ions $(AlOH)(H_2O)^{2+}$, tend to combine through bridging OH groups

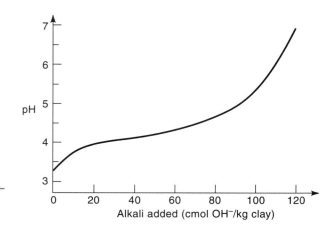

Figure 4.10   pH titration curve for an Al-vermiculite.

and build into polymeric units of six or more Al atoms. The residual H^+ ions associated with these hydroxy-Al ions add to the soil's capacity to neutralize OH^- ions. This acidity, together with the soil's exchangeable acidity ($Al^{3+} + H^+$), is called the *titratable acidity* (or total acidity).

The ability of a soil to neutralize OH^- ions is a measure of its *pH buffering capacity*. The release of H^+ through Al hydrolysis and polymerization augments the soil's pH buffering capacity, with neutralization proceeding until $Al(OH)_3$ precipitates. Similarly, dissociation of H^+ ions from organic matter adds to the buffering capacity. The gradual neutralization of an acid clay by slow addition of an alkaline solution creates a pH titration curve, an example of which is shown in figure 4.10. The flatter the slope of the titration curve, the greater is the buffering capacity of the clay. The strongest acid groups are neutralized first, followed by progressively weaker groups. Some "weak" acidity due to H^+ ion release from hydroxy-Al polymers continues to be neutralized at pH >7.5. The quantity of acid neutralized when a soil is titrated with 0.1M $BaCl_2$/triethanolamine solution at pH 8.2 is a measure of its titratable acidity (cmol H^+/kg), which is invariably greater than the soil's exchangeable acidity. In practice, titration of a soil to a specific pH is used to estimate a soil's *lime requirement* (section 5.5.3).

4.7   *The Supply of Ions to Plant Roots*

4.7.1   *Transport Processes*

Ions move to a plant root by *mass flow* and *diffusion*. Mass flow occurs when the plant is transpiring and absorbing soil water that contains dissolved ions. The ions are swept along in the "transpiration stream" to the root surface. A similar mass flow (sometimes called convective flow) occurs when ions in the soil solution are leached downward by percolating water. Ion diffusion, on the other hand, is a molecular process that occurs in response to a concentration gradient (as discussed in box 3.7) and is independent of the flow of water. Compared to mass flow, diffusion is effective only over relatively short distances.

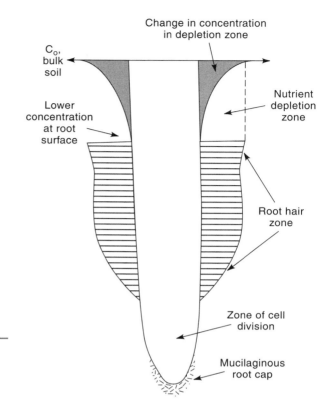

Figure 4.11 Diagram of a grapevine root showing root hairs and the nutrient depletion zone.

4.7.2   *The Absorbing Root*

The rate of nutrient ion uptake increases approximately linearly with the ion's concentration in solution and levels off at high concentrations. The ion concentration at the root surface falls when the uptake rate exceeds the rate of replenishment at the root surface. A "depletion zone" develops around the root. For plants with root hairs, the depletion zone is a cylinder with the internal surface of the cylinder at the root surface and the outside surface at the tips of the root hairs. For grapevine roots, there is a zone of root hairs ca. 20 mm long, some 2 mm back from the root tip (fig. 4.11). Each root hair is up to 0.2 mm long, and there can be 300–400 hairs per mm^2 of root surface (Richards 1983).

Depletion zones are most common with ions that are strongly adsorbed or form insoluble compounds, so their concentration in solution is low. The contribution of mass flow in ion transport to the root is therefore small. Phosphate is a good example. Similarly, ions such as K^+, Fe^{3+}, Zn^{2+}, and Cu^{2+} may show depletion zones around absorbing roots. For these ions, the rate of diffusion across the depletion zone is very important in determining the rate of supply to the root; the effective diffusion rate being inversely related to the sorption capacity of the soil.

For ions such as Ca^{2+}, Mg^{2+}, NO_3^-, and SO_4^{2-}, however, the concentration in the soil solution is normally high enough for mass flow alone to meet the root demand, and no depletion zone develops. For grapevines, the rate of N uptake has frequently been found to vary linearly with the rate of water uptake (Mullins et al. 1992). Many herbicides, which are relatively small organic molecules, are also transported by mass flow and absorbed at about the same rate as the vine's transpiration rate.

4.7.3   *Mechanisms for Enhancing Nutrient Supply and Uptake*

Any mechanism that increases a nutrient's concentration at the root surface should enhance uptake of that nutrient by the plant. Some of the possible mechanisms follow.

4.7.3.1   *Changes in Ion Concentrations in the Soil Solution*

Change in pH. If the uptake of cations relative to anions creates an imbalance in the charge crossing a root surface, either H^+ or HCO_3^- ions will be excreted from the root to maintain charge balance. When N is taken up as NO_3^-, the total of anion charges absorbed usually exceeds that of cations, and HCO_3^- is excreted. HCO_3^- breaks down to CO_2 and OH^-, so the pH in the rhizosphere tends to rise. Only when N is absorbed as NH_4^+ is the pH in the rhizosphere likely to fall. These pH changes are usually about ± 0.4–0.6 pH units and are confined to a few cm of root length at the apex. An increase in pH will lower the availability of micronutrients (except Mo) and may have an effect on P availability, depending on the form of nonlabile P in the soil (section 5.4.2.1). CO_2 is also produced by respiration of the root and rhizosphere microorganisms, but this diffuses away from the root rapidly and has little effect on rhizosphere pH.

Organic exudates. Soluble, low molecular weight C compounds exude from roots into the rhizosphere. Less than half of the exudate comprises organic acids, such as acetic, oxalic, and citric acids, the anions of which can desorb phosphate by ligand exchange (section 4.5.4). Similarly, organic acids may be produced by rhizosphere bacteria. But these and other C compounds are readily used as substrate for growth by competing organisms in the rhizosphere, and it is uncertain whether their concentrations are sufficient to desorb significant amounts of P. Microorganisms also compete strongly with plant roots for any extra soluble P released in the rhizosphere.

Organic acids, such as citric acid, can complex metal cations such as Fe^{3+}, Cu^{2+}, and Zn^{2+}. This chelation process increases the effective concentration of the metal ion in solution near the root surface, which is important in alkaline soils where the concentration of free Fe, for example, is very low because of the insolubility of $Fe(OH)_3$. More research is necessary to determine quantitatively the extent of chelation effects on metal ion uptake by plants, and grapevines in particular.

Solubilization by microorganisms. Several species of bacteria have been grown in pure culture where the only source of P was an insoluble Ca phosphate. Other bacteria obtain P for growth by hydrolyzing organic P compounds, such as the phytates (section 4.4.1). However, if such bacteria grow in the rhizosphere, there is as yet no clear evidence that P taken up by the bacteria is made available in significant amounts to the plant.

4.7.3.2   *Mycorrhizae*

Mycorrhizae are a *symbiotic association* between a fungus and the roots of a higher plant (the host); they have been found to enhance nutrient uptake by a number of plant species. There are two main types of mycorrhizae:

1. *Ectomycorrhizae.* The fungal mycelium forms a sheath—the *Hartig net*—around the root cylinder and also grows through the intercellular spaces in the root. Ectomycorrhizae are primarily associated with woody trees such as pines, oak, beech, birch, alder, poplar, and *Eucalyptus.*
2. *Endomycorrhizae.* This is the form that infects grapevine roots. The fungus does not develop an external sheath, but grows mainly within the root, forming two characteristic structures: the *vesicles*, which are temporary storage bodies in the intercellular spaces, and the *arbuscules*, which are branched hyphae that invade the host cells (fig. 4.12). Hence the common name is *vesicular-arbuscular mycorrhiza (VAM)*. The fungus supplies nutrients to the host and receives a supply of organic substrates for growth from the host.

VAMs are widely distributed on *Vitis vinifera* roots in Australia and have also been recorded in France, Germany, Poland, and Russia (Possingham and Groot Obbink 1971). Although the fungal spores may be present in the soil, they germinate only when a suitable host root is near. If the young mycelium does not penetrate a host root, it dies. Generally, mycorrhizal infection improves the growth of plants, including grapevines, especially on infertile soils, where the mycorrhizal infection enhances the uptake of P. The fungus may mineralize organic P compounds, and the external hyphae provide a pathway for rapid transport of P into the plant. The incidence of infection declines when P fertilizer is applied to the soil, although the extent of this decline varies for cultivated crops. There is evidence for improved water uptake by mycorrhizal plants, probably because of the external hyphal network. *VAMs* can also enhance the uptake of Cu and Zn from soil.

Figure 4.12 Diagram of a vesicular-arbuscular mycorrhiza associated with a plant root (White 1997). Reproduced with permission of Blackwell Science Ltd.

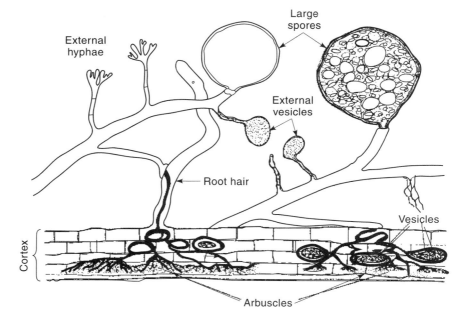

4.8   *Summary Points*

This chapter covered all aspects of how the soil supplies nutrients to plants. The main points are summarized here.

▪ The elements C, H, O, N, P, S, Ca, Mg, K, and Cl are essential for plant growth and reproduction. They are called *macronutrients* because they occur in concentrations >1000 mg/kg of plant DM. The elements Fe, Mn, Zn, Cu, B, and Mo, which are also essential, normally occur at concentrations <100 mg/kg DM, and therefore are called *micronutrients*. Elements in concentrations <1000 mg/kg in the soil are called *trace elements*. They include Cr, Se, I, and Co, which are essential for animals only. Others such as Li, Be, As, Hg, Cd, Pb, and Ni are not required by plants or animals, but are toxic at concentrations greater than a few mg/kg.

▪ Nutrients in soil–plant–animal systems are continuously cycling among a *biomass store* (living plants and animals), an *organic store* (dead plant and animal matter), and an *inorganic store* (minerals and the soil solution). Much of the biomass store is aboveground. The organic store is concentrated in the top 15–25 cm of soil, whereas the inorganic store is distributed through the soil profile. Within each store, nutrients can be transformed from one chemical form to another.

▪ Plants absorb nutrients from a *labile pool* (part of the inorganic store), consisting of ions in solution and adsorbed by clays, sesquioxides, and organic matter. These are called "available nutrients." Nutrients and potentially toxic elements held in insoluble precipitates or strongly adsorbed complexes (organic or inorganic) are *nonlabile*. This subdivision is particularly important for elements such as P, Fe, Mn, Zn, Cu, and Mo.

▪ Ninety-eight percent or more of soil N is in the organic store, but it is converted by *ammonification* and *nitrification* (collectively called *mineralization*) to the mineral forms NH_4^+ and NO_3^-, respectively. The reverse process—incorporation of mineral N into microbial protein—is called *immobilization*.

▪ A minority of plants form symbiotic associations with microorganisms and reduce N_2 gas to NH_3 (incorporated into proteins) within their tissues. This process is called *biological nitrogen fixation*. The most important symbioses of this kind involve legumes and the bacterium *Rhizobium*. Some free-living microorganisms also fix atmospheric N_2. Nitrogen fixation by legume cover crops can provide an important input of N to the biomass store in vineyard soils.

▪ Cations such as Ca^{2+}, Mg^{2+}, K^+, Na^+, and NH_4^+ are held as *exchangeable cations* by negatively charged clays and organic matter. Anions such as $H_2PO_4^-$, SO_4^{2-}, NO_3^-, and Cl^- are attracted to positively charged sites on the edge faces of clays (at pH <6) and sesquioxides (at pH <8). Phosphate ions, which have a high chemical affinity for Al and Fe, are also adsorbed on sesquioxide surfaces by *ligand exchange*. Ions such as $H_2PO_4^-$ and MoO_4^{2-} that attach to surface metal atoms by ligand exchange are *specifically adsorbed* and not easily released.

▪ The combination of fixed negatively charged surfaces and mobile cations in the contiguous solution comprises a *diffuse double layer* (*DDL*) up to 10–20 nm thick. In detail, the *DDL* consists of a Stern layer of cations tightly adsorbed to the surface plus a diffuse layer in which cations are accumulated and anions are in deficit, relative to the bulk solution. The thickness of the *DDL* decreases with the charge on the adsorbed cation and with an increase in ion concentration of the bulk solution. An increase in *DDL* thickness causes flocculated domains of clay crystals to swell in water and may lead to deflocculation.

▨ Cations in the Stern layer are held more strongly because of specific forces, in addition to the simple electrostatic force of attraction. Within the monovalent cation series, the overall strength of adsorption increases in the order $Li^+ < Na^+ < K^+ < Rb^+ < Cs^+$. Similarly, the overall strength of adsorption of the divalent cations increases in the order $Mg^{2+} < Ca^{2+} < Sr^{2+} < Ba^{2+}$.

▨ Sesquioxide surfaces and the edge faces of kaolinite crystals develop pH-dependent charges as a result of the association, or dissociation, of H^+ ions. The attraction of cations and anions is modified because the net surface charge changes with pH. The pH at which the diffuse layer charge is zero defines the *point of zero charge* (*PZC*) of the surface.

▨ The *cation exchange capacity* (*CEC*) is measured by replacement of the resident exchangeable cations with an "index" cation. One common method uses a strong salt solution, for example, M NH_4Cl at pH 7. Others use an index cation of high affinity for the surface at a lower concentration (e.g., 0.1M $BaCl_2$) at the soil pH. The former method overestimates *CEC* in acid soils with a pH-dependent charge; the latter methods are preferred for such soils because they measure the *effective CEC* (*ECEC*). *CEC* values range from 3–20 cmol charge (+)/kg for kaolinites to 100–150 cmols charge (+)/kg for vermiculites.

▨ The difference between the *CEC* and the sum of exchangeable cations (Σ Ca, Mg, K, Na) defines the *exchangeable acidity* (Al^{3+} and H^+). Al^{3+} is an acidic cation because it readily hydrolyzes in water to release H^+ ions. Because of slow clay mineral dissolution, acid clays are primarily Al^{3+}-clays. Neutralization of the hydrolysis products of Al^{3+} (hydroxy-Al ions and polymers) accounts for much of a soil's *titratable acidity* (or total acidity) over the pH range 4–7.5. Titrating a clay or soil with an alkaline solution is a way of measuring its *pH buffering capacity*.

▨ Ions move to plant roots by *mass flow* and *diffusion*. The former process is important for ions in relatively high concentration in the soil solution (e.g., Ca^{2+} and NO_3^-). The latter is the main process for the movement of adsorbed ions such as K^+ and $H_2PO_4^-$, normally present at a low concentration in solution. Many plants, including grapevines, have evolved a fungus–root symbiotic association, or *vesicular-arbuscular mycorrhiza*, that enhances the uptake of P, Cu, and Zn from soils deficient in these elements. Insoluble P compounds may also be dissolved by organic acids secreted by roots and rhizosphere microorganisms.

5   Nutrients for Healthy Vines and Good Wines

5.1   *Soil Fertility and Productivity*

The *fertility* of a soil refers to its nutrient supplying power. It is one of the most important soil factors affecting vineyard productivity, which is measured in tonnes of grapes per ha (or sometimes tons per acre). For viticulture, soil physical properties, notably structure, aeration, and drainage are also very important determinants of productivity, as discussed in chapters 3, 6, and 7. Because vines are grown in permanent rows, and there are many cultural operations, soil physical problems are often more difficult to ameliorate than problems of soil fertility.

Soil fertility is assessed either by observing the condition of vines growing on a particular soil or by measuring the nutrient supplying power of the soil itself. The assessment should include recommendations on how to correct any problems identified. Thus, assessment of soil fertility can be made in two parts:

1. *Diagnosis of nutrient deficiencies or excesses.* The aim here is to identify which nutrients are deficient or in excess and the degree of deficiency or excess. An excess of a nutrient, which may create an imbalance with other nutrients, often leads to a nutrient toxicity.
2. *Estimation of nutrient requirements.* The goal here is to estimate how much of a limiting nutrient is required to achieve optimum growth or how to remedy a toxicity problem. Nutrient amendments can be made with fertilizers, manures, and composts, or by growing cover crops that include legumes.

5.2   *Diagnosis of Nutrient Deficiency*

5.2.1   *The Plant*

Visual symptoms are the signs that indicate a deficiency or excess of one or more essential elements in a plant. In the case of grapevines, such symptoms include

Table 5.1   ***Examples of Element Deficiencies and Their Visual Symptoms in Grapevines***

Deficient Element	Description of Symptoms
N	Overall reduction in growth; leaf yellowing
Fe, Mg, Mn	Interveinal chlorosis; chlorophyll retained along the leaf veins only
K	Starts as a yellowing of older leaf margins. As deficiency worsens, margins die (necrosis) and curl, and the chlorotic areas become a bronze color.
Zn	Stunted lateral shoots with small leaves; fruit set is affected
Cu	Dark green, small leaves; short internodes, stunted growth
B	Stunting of the shoot and death of the shoot tip; leaves near the tip develop interveinal chlorosis and die

chlorosis, stunted growth of shoots, necrosis of leaf margins, irregular fruit set, and small berries. Chlorosis is a generic term for leaf yellowing due to loss of chlorophyll. N deficiency typically causes an overall chlorosis of the leaves, but in other cases chlorosis occurs between the leaf veins (interveinal chlorosis). Some examples of visual symptoms are given in table 5.1 and figure 5.1.

The location of visual symptoms depends on the mobility of the element in the plant tissues, as indicated in table 5.2. Elements are mobile because they are primarily in ionic form in the tissue (e.g., K^+, Na^+, and Cl^-) or because they are in high demand for growth, and so are translocated from older, mature organs to young growing tissues (e.g., N and P). Further details of visual symptoms of deficiency or toxicity are given in specialist books such as Christensen et al. (1978) and Goldspink (1996).

Diagnosing deficiencies and toxicities can be confusing because of the similarity between symptoms caused by different elements (e.g., Fe, Mg, and Mn). The confusion is compounded if more than one element is deficient. Also, visual symptoms appear only after the plant has suffered a check to growth due to a "hidden hunger" for the element. This is illustrated by the well-established relationship between yield and the concentration of an essential element in a plant's tissues (fig. 5.2). The relationship shows that plant yield responds to an increase in the supply of a nutrient, such as N, up to a maximum value. The N concentration in the tissue also increases, and the approximately linear range between "severe deficiency" and "optimum" is used to assess the degree of deficiency. This is the basis for the diagnosis of nutrient status by plant analysis, or *tissue testing*.

The *critical value* is the element concentration in the plant below which an increase in supply leads to increased yield (fig. 5.2). Note that in the "luxury consumption" range, yield may decline due to nutrient imbalances or outright toxicity. A good example in grapevines is toxicity that results from excess concentrations of Na^+ and Cl^- in the leaves, which shows up as severe marginal leaf burn. Perennial crops such as grapevines are more suited to tissue testing than annual crops, for which soil analysis or *soil testing* is more common. However, soil testing is the only tool available to a grower at the establishment stage of a vineyard. Soil testing is also most useful for determining whether lime is needed (section 5.5.3) and for assessing soil salinity (section 7.2.2). It can also be used as a backup for tissue testing. Soil testing is discussed in section 5.3.

(a)   (b)
(c)   (d)

Figure 5.1   Visual symptoms of nutrient deficiency in vines. (a) P deficiency: leaves discolored by reddish-brown blotches in the interveinal area. (b) K deficiency: leaves normal green except for pronounced necrosis extending in from the margins of older leaves. (c) Fe deficiency: leaves overall pale green with extensive interveinal chlorosis. (d) Zn deficiency: leaves dark green except for sharply defined interveinal chlorosis; some shoot stunting. Photographs courtesy of Scholefield Robinson Horticultural Services, Netherby, South Australia. See color insert.

5.2.2   *Sampling Grapevines for Tissue Testing*

Time of Sampling. The concentration of most nutrient elements is highest in young tissues and declines with time. This is especially true of mobile elements such as N, P and K (table 5.2). Therefore, the time of sampling for tissue testing must be specified. Normally, leaves are collected at full flowering or bloom (box 5.1). For trouble-shooting, leaves can be collected at any time to try to confirm

Table 5.2   *Mobility of Elements in Plants and the Location of the First Symptoms of Deficiency*

Element	Mobility	Where Deficiency Symptoms Are Likely to Appear First
N, P, K, Mg, Na, and Cl	Very good	Old leaves
Mn	Variable	Young and midstem leaves
S, Fe, Cu, Zn, and Mo	Very little	Young leaves
Ca, B	Very little to none	Young leaves and growing tip

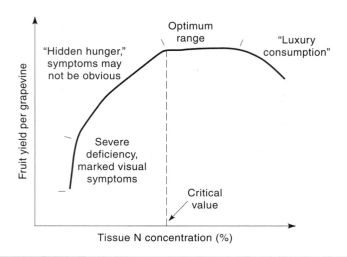

Figure 5.2   Relationship between yield and nutrient concentration in plant tissue.

a deficiency or toxicity suggested by visual symptoms. Midsummer sampling (at veraison) is also used to check on any doubtful results from an earlier sampling. Leaves must not be sampled soon after any foliar spray has been applied.

Plant Part. Because remobilization of elements is possible within the plant, the part to be sampled must be specified. A leaf consists of a blade, or lamina, and a petiole. Leaves are plucked from opposite flower bunches near the base of a shoot, about the fourth or fifth leaf from the base (referred to as "basal leaves"),

---

| Box 5.1 | *Morphological and Physiological Stages of Shoot Growth and Grape Ripening* |

Bud burst: Bud dormancy is broken and young leaves begin to emerge and expand in spring (early October or early April in cool climate regions of the Southern and Northern Hemispheres, respectively).

Flowering (or bloom in North America): 50% or more of the grape bunches or clusters are in flower; can extend over several weeks in cool climates; cell division begins in the fertilized flower.

Fruit set: 50% or more of the bunches have berries that are 3–5 mm in diameter; all unfertilized berries should have been shed. From fruit set to veraison, berry growth is mainly by cell division.

Veraison: The berries begin to soften and change color—a sign of ripening. Midveraison is when 50% of the bunches are coloring (for red grapes). After veraison (postveraison), berry growth is mainly by cell expansion (through water and solute accumulation).

Maturity: Full ripeness attained; bunches are harvested.

Post-harvest: Vines continue to grow, especially through root growth, provided temperature and soil water are adequate.

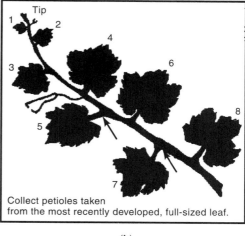

(a)	(b)

Figure 5.3 (a) Position of leaf petioles to be sampled for tissue analysis at flowering (bloom) (b) from mature leaves at veraison (Christensen et al. 1978). Reproduced with permission of Division of Agriculture and Natural Resources, University of California.

as shown in figure 5.3. For later sampling, select the most recently matured leaf on a shoot after the flowering period—usually the fifth to seventh leaf behind the tip. Petioles are preferred over blades, for ease of handling, so the petiole is separated from the blade and placed in a paper bag, to be sent for analysis. If toxicity is suspected, blades may also be sent for analysis because elements such as B accumulate more in the blade than in the petiole. The samples should be rinsed in distilled water or rainwater to remove dust and possible spray contaminants, and blotted dry.

A Representative Sample. Take one leaf per vine from the same variety on each soil type. Usually 75–100 petioles from basal leaves or 100–150 petioles from the youngest mature leaves are adequate for analysis. For comparative purposes, it is wise to take separate samples from areas that are either performing well or poorly.

Test Results. The results are reported in mg element (e.g., K) per kg dry matter, or as a percentage of dry matter (e.g., %N), as described in box 4.1. Remember that biological material is naturally variable, so each mean or average value will have an associated random error. With careful sampling and analysis, this error can be kept to around ±10%. Regular testing over several seasons should show the general trend in nutrient concentrations, even though there may be some random noise in the data. This is the basis of *crop logging*.

Keeping Records. The condition of the vines and soil (especially soil moisture) should be recorded at sampling, as well as the history of fertilizer or manure applications and any cultural operations (e.g., spraying with fungicides). These records, together with the tissue tests and any soil tests, provide a valuable crop log when compiled over several seasons.

Table 5.3   *Macronutrient Concentrations (%) in Petioles for Assessing the Nutrient Status of Grapevines*[a]

Element	Very Deficient	Deficient	Low to Marginal	Adequate	High	Excessive
N (total)		<0.7	0.7–0.84	0.85–1.2	>1.2	
P	<0.15	0.15–0.19	0.2–0.29	0.3–0.5	>0.5	
P (Pinot Noir)	<0.12	0.12–0.14	0.15–0.19	0.2–0.39	>0.4	
K (with adequate N)		<1.0	1.0–1.7	1.8–2.0	2.1–4.4	>4.5
K (vines on Ramsey rootstock)		<3.0		3–4.5		
Ca			<1.0	1.0–2.5		
Mg		<0.3	0.3–0.49	0.5–0.8	>0.9	
			(mg/kg)			
N as NO_3^-	<50	<350	350–600	601–1200	1200–2000	>2000

[a]*Source:* Compiled from Kliewer (1991), Goldspink (1996), Robinson et al. (1997), and Jackson (2000)

5.2.3   *Examples of Tissue-Testing Results*

Critical values for each element are necessary to interpret tissue analyses. These should be determined when no other nutrient is limiting growth and water supply is adequate. The critical values for plants under water stress tend to be lower than for well-watered plants. If these conditions are met, the critical value should be independent of soil type. Grapevine varieties differ to some extent in their uptake of nutrients, which is also influenced by whether the variety is on its "own roots" or a rootstock. However, because knowledge of the critical values for individual varieties is incomplete, the examples given here are generalizations, based on results for Thompson seedless grapes, unless otherwise stated. Tables 5.3 and 5.4 give petiole concentrations for some of the macronutrients and micronutrients, respectively. The semiquantitative interpretations follow the principles of figure 5.2, that is, the critical value for the element is at the lower end of the optimum range. In the case of nitrogen, analyses for NO_3-N may also be used as an indicator of the vine's N status.

Table 5.4   *Micronutrient Concentrations in Petioles for Assessing the Nutrient Status of Grapevines*[a]

Element	Deficient	Low to marginal	Adequate	High	Excessive
Fe			>30		
Cu	<3	3–5	6–30	>40[b]	
Zn	<15	15–25	25–150	>450[b]	
Mn	<25	25–29	30–60		>500
B	<25	25–34	35–70	71–100	>100

[a]*Source:* Compiled from Goldspink (1996), Robinson et al. (1997), and Jackson (2000)
[b]Possible contamination from fungicidal sprays
Note: All measurements in mg/kg.

5.3   *The Soil*

The basis of *soil testing* is to measure that fraction of the total nutrient in a soil that is *available* for plant growth (section 4.2.1). However, the concept of nutrient availability is more complex than at first appears because

- Plants take up nutrient ions from the soil solution in the root zone, in which the total amount of nutrient present is usually insufficient to meet the plant's demand.
- The concentration of a nutrient in the soil solution is buffered by the labile pool of the nutrient (section 4.4.3), the greater part of which comprises the easily desorbable ions on surfaces (for Ca, Mg. K, P, and S). In the case of N, S, and P, a readily mineralizable organic fraction may also be included.
- The rate of replenishment of a nutrient in solution at a root surface depends not only on the soil's buffering capacity for that nutrient, but also on the rate of movement to the root surface by mass flow and diffusion (section 4.7.1).

5.3.1   *Nutrients Held Mainly in an Inorganic Form*

Although these three "availability factors" interact to determine the supply of a nutrient, typically a soil test method involves using a single chemical extractant to measure the amount of a particular nutrient available. A number of acid or alkali extractants have been advocated for available P, for example, but one of the more widely used is a solution of $0.5M$ $NaHCO_3$ buffered at pH 8.5. For Ca, Mg, and K, the amount of exchangeable cations, as measured by methods described in box 4.5, can be used. For SO_4-S, a solution of $0.01M$ $Ca(H_2PO_4)_2$ is sometimes used, and for the micronutrients Fe, Zn, Cu, and Mn, a solution of $0.05M$ or $0.1M$ EDTA or DTPA (both metal complexing agents) are used. Further details on soil testing are given in specialized texts, such as Rayment and Higginson (1992) or Jones (1999).

Soil analyses are carried out on a composite sample of soil made up from individual soil cores, usually taken to a depth of 15 cm, in a way that covers the range of variability in a vineyard. In the case of N as NO_3^-, samples may be taken to depth in the profile (e.g., to 50 cm). Subsamples from individual soil types should be mixed separately to provide composite samples for analysis.

5.3.2   *Nutrients Held Mainly in Organic Form*

Nitrogen is the main element in this category. Grapevines absorb N in the form of NH_4^+ and NO_3^- ions, but the latter predominate in most soils because of nitrification. The availability of soil N therefore depends on the amount of mineral N in the soil, and the rate of mineralization of organic N.

Ammonium is held as an exchangeable cation that is easily displaced into the soil solution where it moves with NO_3^-, by mass flow to the roots of a transpiring plant. Both ions also move from regions of higher to lower concentration by diffusion. Thus, all the *mineral N* (except for any nonexchangeable NH_4^+ held in micaceous clay lattices) is available to the plant. Mineral N is measured by extracting the soil in a solution of 2M KCl and by calculating the amount in mg N/kg soil.

The balance between N mineralization and immobilization during decomposition of organic matter depends on the C:N ratio of the substrate (section 2.3.1.2). The outcome of this "balancing" is the *net* mineralization rate, given by the equation

$$\frac{dN}{dt} = -kN \qquad (5.1)$$

where N is the amount of organic N per unit soil volume, t is time, and k is a decay coefficient (compare equation 2.1). The expression $\frac{dN}{dt}$ is the *instantaneous* rate of net mineralization. The way in which this equation is used to obtain values of mineralized N ($N_{min.}$) over a period of time is illustrated in appendix 4. The mineral N first appears as NH_4^+, but this is usually rapidly converted to NO_3^- by nitrifying bacteria.

The decay coefficient k is an *averaged* parameter, because there are differences in net mineralization rate among fresh residues (grass mulch and green manures), animal excreta (chicken manure), and resident soil organic matter (*SOM*). For example, fresh residues of C:N ratio <25 decompose and mineralize N much faster than *SOM*, which has achieved a high degree of stability after many cycles of synthesis and decomposition. Thus, $N_{min.}$ values will vary according to the following factors:

- the amount of residues and organic manures returned to the soil, relative to the *SOM*,
- the C:N ratio of the residues and manure, and their distribution in the soil (which affects their accessibility to microorganisms),
- soil management (e.g., soil pH, and whether the soil is cultivated or drained), and
- environmental factors (especially temperature and effective rainfall).

Cover crops. Cereals, grasses, and legumes, sometimes in combination, are often grown in the inter-rows of vineyards to protect the soil from erosion and compaction, especially during winter (section 7.3.1). When cultivated into the soil in early summer, cover crops containing legumes provide young green material of low C:N ratio that decomposes quickly and is an important source of mineral N to the vines. This is called *green manuring*. Green manure crops that decompose quickly have little effect on *SOM*; straw and bark mulches, on the other hand, have high C:N ratios and are slow to decompose. Such mulches can increase *SOM*, but during decomposition, they "lock up" soil mineral N and make it unavailable to the vines. Eventually, as the C:N ratio of the mulch residues slowly declines, some of this immobilized N is mineralized.

Values of $N_{min.}$ for a range of organic materials and soil conditions are given in table 5.5.

5.3.3   *Calibration of Soil and Plant Tests*

Irrespective of which soil extraction or tissue analysis method is used, the test must be calibrated against a measure of plant performance, usually the yield of grapes

Table 5.5   *Estimates of* N_{min} *During One Year from Net Mineralization of Soil Organic Matter and Green Manure*

Soil Organic Matter (50% C, C:N Ratio = 10) *SOM* (t/ha)	Organic N Content, N_o (kg N/ha)	N_{min} (kg/ha)[a]		
		k Value (1/yr)[b]		
		0.002	0.005	0.01
30	1500	3	7.5	15
40	2000	4	10	20
50	2500	5	12.5	25
Green Manure (with Legume) (40% C, C:N Ratio = 20) DM (t/ha)	Organic N Input, N_o (kg N/ha)	N_{min} (kg/ha)[c]		
		k Value (1/yr)[b]		
		0.5	0.75	1.0
1	20	8	10	13
1.5	30	12	16	19
2	40	16	21	25

[a]Calculated from equation A4.8 in appendix 4

[b]The decay coefficients vary with the type of soil and its management, as well as the type of green manure crop.

[c]Calculated from equation A4.4 in appendix 4

per vine or per ha. Calibration requires a number of trials at different sites in which the response of vines to increasing amounts of a specific nutrient is measured. It is important that other nutrients are not limiting and the vines do not suffer serious water stress. These tests are commonly called *fertilizer rate trials*. Ideally, the results should be specific to a particular variety, but they can cover a range of soil types. The test nutrient can be applied as a commercial fertilizer, such as urea to supply N, or muriate of potash (KCl) to supply K.

The yield response to nutrient addition usually follows the law of diminishing returns, which states that the yield increase for successive equal increments of a nutrient becomes steadily less. This relationship is described by the equation

$$y = A - B \exp(-Cx) \qquad (5.2)$$

where y is the yield per ha and x is the amount of nutrient applied in kg/ha. Also, A is the maximum yield, $(A - B)$ the yield from the soil without added nutrient (the control or "check" yield), and C the rate of change of y with x. Figure 5.4a shows an example of a yield response curve to N fertilizer. Because A is approached asymptotically, the practical maximum is often set at 90% of the true maximum (Goldspink 1996). If the plant tissue is analyzed, the *critical value* for the tissue test would be determined at 90% of maximum yield. Similarly, if the soil is analyzed, the critical value for the soil test is determined in the same way.

Figure 5.4b has the same response curve plotted to show the increasing value of the product relative to the cost of fertilizer applied. When the value of the extra yield from fertilizer (curve AB) is compared with the fertilizer cost line (AC), the vertical XY indicates the fertilizer rate that gives the biggest profit from fertilizer. If the fertilizer cost changes, XY moves to the right or left to a new point of maximum profit. Similarly, if the crop value changes due to quality factors, for example, the curve AB will change and the position of XY moves.

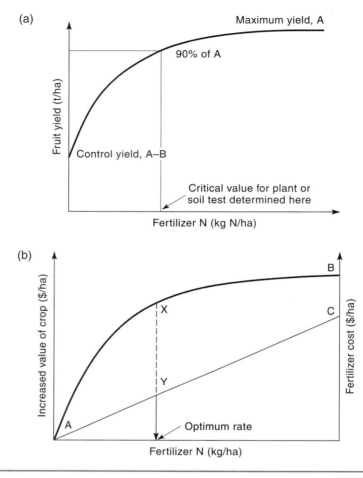

Figure 5.4 (a) Generalized yield response to N fertilizer (b) Relationship between product values and fertilizer cost showing optimum yield.

These figures clearly demonstrate the distinction between *maximum yield*, where further additions of fertilizer do not produce an increase in yield (and yield may even start to decrease due to a nutrient imbalance or toxicity), and *optimum yield*, where the increase in product value relative to the cost of additional fertilizer is at a maximum. The concept of optimum yield is particularly relevant to the wine and grape industry because high yields per ha may be produced at the expense of fruit quality. This is especially true of N fertilization, and explains why vignerons aim to restrict the supply of N to vines at critical stages of growth (section 5.4.1.4).

5.3.4   *Nutrient Requirement Converted to Fertilizer Requirement*

A nutrient requirement, as assessed from a response curve such as in figure 5.4, must be converted into a fertilizer requirement. Choosing a fertilizer depends on

---

Box 5.2   *Determining the Best Fertilizer Option to Meet a Nutrient Requirement*

Suppose the optimal N requirement, determined from figure 5.4b, is 40 kg/ha. The cost per kg N content determines the slope of the fertilizer cost line AC in figure 5.4b. The unit cost is given by

$$\text{Cost/kg N} = \frac{\text{Cost/tonne}}{\text{N content} \times 10} \tag{B5.2.1}$$

If the fertilizers have similar costs per tonne, clearly the most economical source of N is the fertilizer with the highest N content (expressed as a percentage). Usually, the price of urea per tonne is the lowest, so the unit cost is much lower for urea than other N fertilizers commonly used in vineyards (table 5.6). However, losses of N by volatilization can be higher, and the acidifying effect of ammonium oxidation can be a problem with urea (section 5.4.1.2). These factors must be balanced against the economic advantage.

The quantity of fertilizer Q required to supply 40 kg N/ha is given by

$$Q = \frac{40 \times 100}{\text{N content}} \tag{B5.2.2}$$

If the fertilizer chosen is ammonium nitrate (34% N), Q = 118 kg/ha. Because there are 10,000 m² per ha, this is equivalent to 11.8 g/m². If the fertilizer is to be dissolved in irrigation water and applied through drippers—the process of *fertigation*—we must know the quantity of water to be applied per vine (chapter 6). The solubility of the fertilizer then becomes an important factor.

---

the preferred form of fertilizer and its cost. An example of the calculations is given in box 5.2.

5.3.5   *Precision Viticulture*

As with other intensively managed crops, yield and quality can vary markedly within a vineyard due to subtle changes in soil type, soil depth, drainage, and nutrient supply. The spatial pattern of variation may be consistent from year to year, in which case blocks of vines can be identified where soil management and fertilizer practice are tailored to achieve the desired soil physical and chemical conditions. This is the basis of *precision viticulture*.

The spatial pattern of variation is mapped by manual soil surveys or real-time sensing with mobile instruments. The position of each measurement is determined by a Global Positioning System (GPS) and the coordinates entered into a Geographical Information System (GIS), together with the "soil attribute" data (pH, profile depth, soil texture, salt content, and so on). Also, yield monitors are available for mounting on grape harvesters; these monitors allow the spatial pattern of yield variation to be recorded digitally, entered into the GIS, and mapped. An example of such a yield map is shown in figure 5.5 for a block of Ruby Cabernet grown under irrigation in Sunraysia, Victoria. The generally low yields (<9 t/ha) in the northwest sector were due to poor drainage (a shallow impermeable subsoil), whereas the high yields (>20 t/ha) in the eastern part were obtained on a deep, permeable sandy loam. The recommended action is that drainage in the

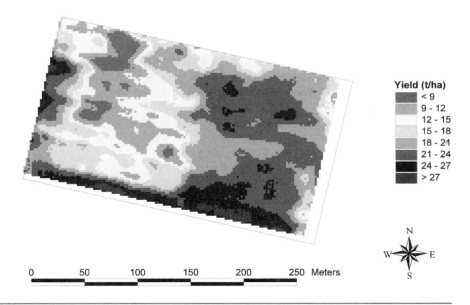

Yield (t/ha)
- < 9
- 9 - 12
- 12 - 15
- 15 - 18
- 18 - 21
- 21 - 24
- 24 - 27
- > 27

0   50   100   150   200   250 Meters

Figure 5.5   Grape yield map for precision viticulture (courtesy of R. G. V. Bramley; see color insert).

northwest sector be improved by installing under-drainage. If successful, this would raise the overall productivity of the block substantially.

The essence of the precision approach to viticulture therefore involves three steps: (1) acquisition of real-time or existing spatially referenced information (yield, soil data, disease incidence, pruning weights, etc); (2) analysis and interpretation of this information using a decision support system (DSS), usually incorporating some modelling of soil–plant interactions; and (3) implementation of site-specific management (relating to fertilizers, irrigation, spraying, fruit thinning, time of harvesting, etc).

As more information is recorded in successive seasons, the vineyard and crop response to environmental factors and management can be updated, the DSS revised, and management actions modified through an iterative process. Precision nutrient management is potentially a big improvement on the conventional practice of taking bulk soil or plant samples for diagnostic testing, where variation within a site is averaged. Targeted nutrient management not only improves overall yields, but also reduces the waste of nutrients such as NO_3^- through leaching and/or denitrification.

5.4   *Inputs and Outputs of the Major Nutrients*

5.4.1   *Nitrogen*

The quantity of N per ha of vines depends on the planting density, as well as the age and vigor of the vines. Data from California (Mullins et al. 1992) indicate N contents (roots, shoots, and fruit) of 100–128 g N per vine for 10-year-old vines at densities of 1120–1680 per ha. This suggests N contents from 112 kg N/ha at

---

Box 5.3	*Atmospheric Inputs of Nitrogen by Processes Other Than N_2 Fixation*

Various oxides of N (mainly NO and NO_2) are formed during the combustion of fossil fuels in motor vehicles. Some are also formed by forest fires and lightning discharges. The oxides react with hydroxyl free radicals in the air to form nitric acid, which contributes to *acid rain*. NH_3 volatilizes from manures, rotting vegetation, and soil, especially after the application of fertilizers such as urea (section 5.4.1.3). Some NH_3 is absorbed through the stomata in plant leaves, while the remainder is dissolved in rain or forms salts, such as ammonium sulfate $(NH_4)_2SO_4$ according to the reaction

$$2NH_3 + H_2SO_4 \longleftrightarrow (NH_4)_2SO_4 \qquad (B5.3.1)$$

Sulfuric acid (H_2SO_4) is formed from S oxides (see box 5.5). $(NH_4)_2SO_4$ forms an aerosol that is deposited on soil and plant surfaces as *dry deposition* (which also includes any NH_3 and SO_2 uptake through leaf stomata). The total N input from the atmosphere is 5–60 kg/ha/yr, depending on the extent of air pollution.

---

the lowest density to >500 kg N/ha at densities of 5000/ha or more. The amount of N removed in the fruit is between 1.2 and 2.4 kg N/t grapes, with the norm being ca. 1.5 kg N/t.

N from the atmosphere enters the soil–plant system in several forms. Fixation of N_2 gas by legume cover crops is discussed in section 4.2.2.1. Other forms—the oxides of N (NO_x)—come down in rain or as dry deposition (box 5.3). The average annual N input from the air in industrialized parts of Europe is 20–30 kg/ha, which can make a significant contribution to the N requirement of vines.

The other major N input to the soil is through fertilizer and manure, although leguminous cover crops are becoming more important, as part of an overall drive to improve soil quality (chapter 7). Manures are discussed in section 5.7, but the main forms of N fertilizers are discussed here.

5.4.1.1 Forms of N Fertilizers

The more important water-soluble N fertilizers used in viticulture are given in table 5.6. Some of these, such as Nitram and urea, supply a single macronutrient (N). Others, such as KNO_3 and $(NH_4)_2SO_4$, supply more than one macronutrient and are called *multinutrient* fertilizers. *Mixed fertilizers*, which may be solid or liquid, are made by mixing single or multinutrient fertilizers, and they are usually identified by their N:P:K ratio. For example, the mixed fertilizer "Horticulture Special" has a composition of 10–3–8, meaning 10% N, 3% P, and 8% K. In the trade, fertilizer analyses are sometimes given as the ratio of $N:P_2O_5:K_2O$. For simplicity, in this book analyses are given as percentages of the elements, unless otherwise stated.

Slow-release fertilizers, such as ureaform (comprising ureaformaldehyde polymers, 21–38% N), IBDU (isobutylidene diurea, 32 % N), and SCU (sulfur-coated urea, 37–40% N), are also available. They are sparingly soluble N compounds, developed to avoid the occurrence of high, localized concentrations of mineral N in the soil, which predispose to N loss (section 5.4.1.2). Slow-release fertilizers include both synthetic and natural *organic fertilizers* (e.g., blood and bone, hoof,

Table 5.6   *Forms of Soluble N Fertilizer Used in Vineyards*

Compound	Formula	N Content (%)	Comments
Potassium nitrate	KNO_3	13.4	KNO_3 and $Ca(NO_3)_2$ are very
Calcium nitrate	$Ca(NO_3)_2$	15.5	soluble, suitable for fertigation; nonacidifying, but more expensive per kg N than other forms
Ammonium sulfate	$(NH_4)_2SO_4$	21	Supplies N and S in soluble forms; acidifying
Ammonium nitrate ("Nitram")	NH_4NO_3	34	NO_3^- is immediately available; NH_4^+ is adsorbed by soil and oxidized to NO_3^-
Urea	$(NH_2)_2CO$	46	Very soluble, high N content; prone to volatilization if not washed into the soil; acidifying
Urea ammonium nitrate solution	$(NH_2)_2CO$, NH_4NO_3 in water	30–32	Has the advantages and disadvantages of urea and NH_4NO_3

and horn meal). Short-term plant response to these fertilizers is usually inferior, per kg of N, compared with the soluble forms; but they are useful on sandy soils, especially if N fertilizer is applied during winter.

The amount of N fertilizer required depends primarily on the balance among N inputs, N removal in prunings and in the fruit at harvest, and any losses. Before discussing how to estimate amounts, we must consider the loss pathways for N.

5.4.1.2   *Nitrification, Leaching, and Acidification*

NH_4^+ ions from N fertilizer, or from net mineralization of organic N, will nitrify in the soil (section 4.3). The overall reaction is described by equation 4.4, which shows that 2 moles of H^+ ion are produced for every mole of NH_4^+ ion oxidized. Thus, adding NH_4-N fertilizers has the potential to acidify the soil. If the NO_3^- formed is absorbed by the vine, there is some compensatory release of HCO_3^- (which breaks down to CO_2 and OH^- ions) by the roots during uptake. Moreover, if there is recycling of plant N to the soil through leaf fall and decomposition, there is no net acidification from this component. But less than 50% of fertilizer N may be taken up by the vines. Of the NO_3-N not taken up, some is immobilized, but much is lost by leaching or denitrification. The NO_3^- leached is accompanied by cations, such as Ca^{2+}, because the cations are displaced by H^+, as shown here:

$$Ca^{2+}\text{-clay} + 2H^+ \leftrightarrow 2H^+\text{-clay} + Ca^{2+} \qquad (5.3)$$

The H-clay becomes an Al-clay (section 4.6.3.1). Potentially, the most acidifying N fertilizers are those in which all the N is present as NH_4-N, such as urea and $(NH_4)_2SO_4$. Less acidifying is NH_4NO_3, and $Ca(NO_3)_2$ and KNO_3 have no acidifying effect at all. This is one of the reasons why $Ca(NO_3)_2$ is favored as an N fertilizer in vineyards, especially when applied by drip irrigation. KNO_3 is not favored if the soil already supplies adequate K.

Urea is very soluble and may leach with irrigation water to some depth before it undergoes hydrolysis in water. The hydrolysis reactions that occur are

$$(NH_2)_2CO + 2H_2O \rightarrow (NH_4)_2CO_3 \qquad (5.4)$$

$$(NH_4)_2CO_3 + 2H_2O \leftrightarrow 2NH_4OH + H_2CO_3 \qquad (5.5)$$

The NH_4OH produced by reaction 5.5 readily dissociates into NH_4^+ and OH^- ions, and the former ions are subsequently oxidized to NO_3^-. Thus, acidification through nitrification can occur at depths of 40–60 cm in soil under drippers, especially if a large "slug" of urea is applied. Acidity at depth is difficult to correct by liming (section 5.5.3).

5.4.1.3 Gaseous Losses of N

The NH_4OH formed by the hydrolysis of urea is unstable, especially at pH >8 when it dissociates and releases OH^- ions. For a period of 1–2 days, the soil pH around a urea granule rises as high as 9–10 and NH_3 volatilizes, according to the reaction

$$NH_4OH \leftrightarrow NH_4^+ + OH^- \leftrightarrow NH_3 \uparrow + H_2O \qquad (5.6)$$

When NH_4^+ breaks down and NH_3 gas escapes, the remaining H^+ ions neutralize the OH^- ions, so the overall process is pH-neutral. NH_4^+ ions that decompose in this way obviously are not oxidized to NO_3^-, and hence do not contribute to soil acidity. The loss of NH_3 is most serious when urea is applied to the soil surface, and there is insufficient rain or irrigation to wash it into the soil. High soil temperatures and wind accelerate NH_3 volatilization, and in extreme cases up to 50% of the N in surface-applied urea can be lost.

Another process of gaseous N loss is *denitrification*. This occurs when part or all of the soil becomes waterlogged, as discussed in section 5.6.1. Because vineyard soils should be well drained, and the fertilizer N inputs are relatively small, denitrification losses should normally be small too.

5.4.1.4 Timing and Amount of N Fertilizer

The dynamics of vine N uptake are determined by shoot demand and root growth. A generalized time trend for shoot and root growth, and N uptake from the soil, in relation to phenological development of the vine, is shown in figure 5.6. Root growth lags behind shoot growth from bud burst to early flowering. During this period, N reserves in the roots and trunk are mobilized to the developing leaves, and there is little uptake from the soil. The peak period of soil N uptake is from full flowering to veraison, which coincides with the period of the most rapid root growth. From veraison to harvest, the bunches are the main accumulators, with most of the N coming from woody tissues and the shoots, and only a minor portion from the soil. After harvest, N is accumulated in the permanent structures, mainly by translocation from leaves and shoots; but up to one-third of the N may be absorbed from the soil, more so in regions with long hot summers where the roots continue to grow after harvest. This stored N provides the reserve for early shoot growth during the following season and has an important influence on the fruitfulness of the vines during that season. Given this pattern of N uptake, in cool regions with significant spring rainfall, N fertilizer must not be applied

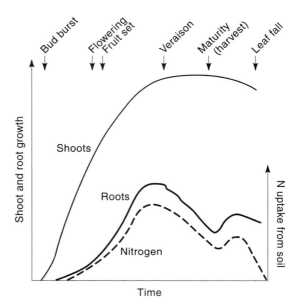

Figure 5.6   Time trend in shoot and root growth and N uptake by grapevines (redrawn from Freeman and Smart 1976, Conradie 1980, and Lohnertz 1991).

before bud burst because of the potential for NO_3^- loss by leaching. The best time for application is from preflowering to preveraison. However, in warm climates, N applied after harvest can be effective because it increases the N reserves that are utilized the following spring.

An adequate supply of N to the vine is most important for shoot growth and photosynthesis, which produces the sucrose subsequently translocated to the fruit. If the N supply is deficient, bud fertility and therefore yield are reduced, and the N concentration of the berries will be too low. The critical measure of berry N is the *Yeast Assimilable Nitrogen* (*YAN*), comprising NH_4-N and amino acid N (excluding proline), which should be in the range 200–300 mg N/L in the juice. If *YAN* is too low (<140 mg/L), the fermentation rate will be too slow and may become "stuck." The wine maker can correct this problem by adding DAP (see table 5.7) to the must. However, if the *YAN* is too high, the fermentation proceeds too rapidly and poor-quality wine is produced, particularly in the case of red grapes. Too high a supply of N to the vine, especially when coupled with plentiful water, can lead to *excess vigor*—a condition of long, straggling shoots and a dense canopy that overshades the bunches and results in lower berry color and flavor. We return to the topic of excess vigor in chapter 6.

Supplying the correct amount of N fertilizer at the right time is important for quality control in grape production and wine making. Examples of how to estimate the amount N fertilizer required are given in box 5.4.

5.4.1.5   *Residual Effects*

A *residual effect* occurs when enough of a fertilizer element applied in one year remains available for growth during the next year. The residual effect of N fertilizer in vineyards depends on the rate and form of N applied, its timing, vine root development, the amount of rain and irrigation, and soil properties. Provided the

Box 5.4   *Estimating the Appropriate Amount of N Fertilizer for Wine Grapes*

The amount of fertilizer required for one season is estimated using an *N balance* approach:

$$\text{Amount of N required} = \text{Inputs} - \text{Crop removal} - \text{Losses} \qquad \text{(B5.4.1)}$$

Some results using equation B5.4.1, assuming a 10 t/ha grape yield, are given in table B5.4.1. The first example is typical of a vineyard in southeastern Australia, for which atmospheric N input is low. The second example is more typical of a northern European vineyard where atmospheric inputs are higher, and the soils, through past fertilizer and manure inputs, are more fertile (Schaller 1991). Inputs from the net mineralization of soil N, and a leguminous green manure crop, can be included by following the procedure outlined in table 5.5. Volatilization of NH_3, even when urea is used, can be minimized through fertigation, but water application must be managed to avoid leaching losses.

In both cases, the amount of fertilizer N required is not large (12–32 kg N/ha/yr). This is consistent with the practice in the Bordeaux region, where ca. 30 kg N/ha/yr are applied, and also for mature vines in the Willamette Valley of Oregon, United States. However, the rate of N input may be two to three times greater on young vines (up to 2 years old), and on sandy soils of low organic matter content. In all cases, the grower should be prepared to adjust the fertilizer input on the basis of tissue testing for N (section 5.2.1).

Table B5.4.1   *Examples of N Fertilizer Required for a Crop of 10 t/ha, Calculated from an N Balance*

Example 1

Atmospheric Inputs	Net Soil Mineralization	Green Manure Crop	Crop Requirement (Shoots Plus Fruit) Less Recycling[a] (kg N/ha)	Estimated Leaching Plus Gas Losses[b]	Amount of N Required as Fertilizer
10	10	0	37.5	10	27.5
10	10	0	37.5	15	32.5
10	10	20[c]	37.5	25	22.5

Example 2

Atmospheric Inputs	Net Soil Mineralization	Green Manure Crop	Crop Requirement (Shoots Plus Fruit) Less Recycling[a] (kg N/ha)	Estimated Leaching Plus Gas Losses[b]	Amount of N Required as Fertilizer
25[d]	20	0	37.5	20	12.5
25[d]	20	0	37.5	30	22.5
25[d]	20	20[c]	37.5	50	22.5

[a]Average amount of N required for growth = 75 kg N/ha (range 68–84) (Mullins et al. (1992) less recycling of 50% through leaf fall and prunings
[b]Depending on soil type and surplus rainfall (plus any irrigation); see table 7.6.
[c]See table 5.5.
[d]*Source:* Data from Schaller (1991) for German viticultural areas

amount and timing of fertilizer N are well matched to plant demand, the fraction of fertilizer N remaining in *mineral* form in the soil when the vines shut down in autumn should be small. Any NO_3^- remaining is vulnerable to leaching during the winter. Some of the fertilizer N is also converted into labile organic N forms, which are readily mineralized the following spring. In some German vineyards where fertilizer inputs have been very high, there is evidence of unacceptably high N losses (>100 kg N/ha/yr) by leaching to groundwater (Schaller 1991). The environmental impact of high N losses is discussed in chapter 7.

5.4.2   *Phosphorus and Sulfur*

The amount of P and S removed per tonne of harvested grapes averages about 0.3 and 0.2 kg, respectively. P is important for metabolic activity (ATP synthesis) and healthy bud development; S is important for synthesis of some of the amino acids.

In soils with adequate organic matter (>2% C), mineralization of organic forms of P and S provides a major input, as discussed in section 4.4. P concentrations in rain are very low, and more P is deposited by dry deposition (mainly dust during dry weather). The rate of deposition is ca. 0.1–0.5 kg P/ha/yr. The amount of S deposited from the atmosphere is much higher, from <5 to 70 kg/ha/yr. Atmospheric inputs of S are discussed in box 5.5.

5.4.2.1   *Forms of P Fertilizers*

Phosphate fertilizers consist of water-soluble *orthophosphates* and polymerized orthophosphates, called *polyphosphates*, and water-insoluble *mineral* and *organic phosphates*. The natural rock phosphates, consisting of minerals of the apatite type with $CaCO_3$, SiO_2, and other impurities, are the raw material from which water-soluble P fertilizers are made by treatment with acid. Traditionally, H_2SO_4 was used to produce *single superphosphate* (*SSP*), but now phosphoric acid (H_3PO_4)

Box 5.5   *Forms of S Input from the Atmosphere*

Sulfur is emitted into the air as SO_2 from the burning of fossil fuels. Near coasts, small amounts of methyl sulfides and H_2S are released from marine sediments. Several processes are involved in the deposition of atmospheric S on soil and vegetation:

- SO_2 is directly absorbed through leaf stomata.
- SO_2 is oxidized to sulfur trioxide (SO_3), with the SO_3 reacting with water to form sulfuric acid (H_2SO_4).
- H_2SO_4 dissolves in rain (hence the term *acid rain*), or is neutralized by NH_3 gas to form $(NH_4)_2SO_4$, which is dissolved in rain, or is deposited "dry" (box 5.3). The NH_4^+ ion contributes to soil acidification through nitrification.

Because of the adverse effect of acid deposition on natural ecosystems, industry has been required to reduce S emissions from fossil fuels. Over northwest Europe, for example, total S emissions have fallen by ca. 50% since 1970. Currrently, atmospheric inputs are in the range of 10–30 kg S/ha/yr and are likely to fall to 5–10 kg S/ha/yr. The contribution of S to the total acid deposition from the air has fallen from two-thirds to one-half.

Table 5.7   ***Forms of P Fertilizer Used in Vineyards***

Compound	Formula	P Content (%)	Comments
Single superphosphate (SSP)	$Ca(H_2PO_4)_2.H_2O$ plus $CaSO_4.2H_2O$	9	Made from phosphate rock by reaction with H_2SO_4; 80–90% water soluble
Triple superphosphate (TSP)	$Ca(H_2PO_4)_2.H_2O$ plus traces of S	19–21	Made from phosphate rock by reaction with H_3PO_4; >80% water soluble
Monoammonium phosphate (MAP)	$NH_4H_2PO_4$	21–26	Very high water solubility, supplies N and P; suitable for fertigation
Diammonium phosphate (DAP)	$(NH_4)_2HPO_4$	20–23	High water solubility, supplies N and P; suitable for fertigation
Phosphate rock	$Ca_{10}(PO_4)_6F_2$ with variable SiO_2, $CaCO_3$ and sesquioxide impurities	6–18	Insoluble mineral fluorapatite; P content varies with the source
Bone meal and guano	Mainly apatite	5–13	Bone meal is an animal product; guano is consolidated bird droppings.

is more widely used to produce high-analysis P fertilizers (e.g., *triple superphosphate*, TSP). Forms of P fertilizer and their P contents are given in table 5.7.

Reactions of Water-soluble P Fertilizers in the Soil. The active constituent of SSP and TSP is *monocalcium phosphate* (MCP), with the formula $Ca(H_2PO_4)_2.H_2O$ (table 5.7). When granules of these fertilizers are placed in soil, they take up water until a saturated solution forms. In dry soil, this solution is very acid (pH ~ 1.5) and concentrated in P (ca. 4 mols/L) and Ca (ca. 1.4 mols/L); in wetter soil, the solution is less concentrated. At this very low pH, soil minerals around the granule dissolve, releasing SiO_2, Al, Fe, and Mn ions. These ions, together with Ca and P from the granule, are subsequently precipitated as complex new compounds—the *soil-fertilizer reaction products*—as the pH slowly rises. In moist soil, granules can dissolve in 24–36 hours, leaving the reaction products and a residue of insoluble dicalcium phosphate ($CaHPO_4$) containing about 20% of the original P. The conversion of water-soluble fertilizer P to less soluble compounds in the soil contributes to the process of "P fixation," whereby P becomes less available to plants.

There are two important points to note about SSP and TSP dissolution in soil:

1. Any localized acidity is neutralized by the reaction with soil minerals, so water-soluble P fertilizers are not inherently acidifying.
2. The speed of the reaction means that a plant feeds not so much on the fertilizer itself, but on the fertilizer reaction products. These are metastable compounds that revert slowly to more stable (but less soluble) products, such as amorphous $AlPO_4$ and $FePO_4$ in acid soils or octacalcium phosphate and hydroxyapatite in neutral to alkaline soils.

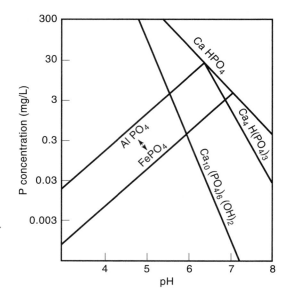

Figure 5.7 Solubility relationships between pH and P for P fertilizer reaction products in the soil (redrawn from White 1997).

As the pH rises, $AlPO_4$ and $FePO_4$ dissolve and hydrolyze to release P. Calcium phosphates, on the other hand, become more soluble as the pH decreases. The overall influence of soil pH on the solubility of P from fertilizer reaction products is shown in figure 5.7. This diagram shows that, in a soil regularly fertilized with P, the soil solution P concentration should be highest between pH 6 and 7.

Reactions of Insoluble P Fertilizers in Soil. Natural rock phosphates are sometimes used as fertilizers in vineyards. In contrast to SSP and TSP, which are granulated, rock phosphates are best applied in powder form to increase the area of contact between the fertilizer and the soil. Phosphate rock (PR) is arbitrarily classified as "reactive" or "unreactive" (hard) on the basis of the PR's solubility in chemical extractants, for example, M ammonium citrate at pH 7 (Australia and the United States), or 2% formic acid (European Union). Hard rocks such as Florida pebble (from the United States) are used for the manufacture of soluble P fertilizers, but the more reactive phosphate rocks (RPRs), such as from North Carolina and North Africa, may be used for direct application. The dissolution of these RPRs depends not only on the chemical composition of the rock itself, but also on soil and climatic conditions. The general reaction in the soil is

$$Ca_{10}(PO_4)_6F_2 + 12H^+ \leftrightarrow 6H_2PO_4^- + 10Ca^{2+} + 2F^- \qquad (5.7)$$

from which we see that the dissolution requires H^+ ions. Because H^+ ions are consumed as RPRs dissolve, there is a small liming effect.

RPRs can be used on grapevines since they are perennial and do not need a rapid uptake of P, except during the establishment phase. However, they are not very effective unless used on acid soils (pH <6 in water) with adequate soil moisture (>800 mm of rain or irrigation, reasonably well distributed through the year). RPRs are acceptable in organic viticulture (section 5.7), but they should be applied to the soil only in a finely divided form.

5.4.2.2   *Forms of S Fertilizers*

Sulfur is the forgotten nutrient in vineyards, mainly because it is a constituent of several fertilizers, such as SSP (11% S), potassium sulfate (K_2SO_4, 18% S), and gypsum ($CaSO_4.2H_2O$, 19% S). A significant amount of sulfur can come from atmospheric input. In addition, elemental S is used as a fungicide. Although up to 28 kg/ha of "wettable S" may be applied in one year, much of this volatilizes. The residue is washed into the soil where it is slowly oxidized by specialist *Thiobacillus* bacteria, according to the reaction

$$S + 3/2O_2 + H_2O \rightarrow 2H^+ + SO_4^{2-} \tag{5.8}$$

Equation 5.8 shows that the oxidation of 32 kg S/ha produces 2 moles H^+ ion/ha. Thus, S applied as a fungicide can cause soil acidification, which must be taken into account in managing the pH of vineyard soils.

5.4.2.3   *Leaching and Residual Effects*

Because of P fixation in the soil, P losses by leaching are usually small (0.1–0.2 kg/ha/yr). Exceptions occur with very sandy soils under high rainfall or irrigation, or with soils that have been heavily fertilized for many years, where losses can amount to 3–5 kg P/ha/yr. Sulfate, on the other hand, is not strongly adsorbed and so is leached from the profile, except in dry climates when it may accumulate in the subsoil as gypsum.

Only a small fraction of fertilizer P is absorbed by vines during the year of application. Much of the remainder is retained as fertilizer reaction products, which become less soluble with time, and the rest is adsorbed to clays and sesquioxides (section 4.5.4). A rough guideline is that two thirds of the soluble P remains available after one year, one-third after two years, one-sixth after three years, and none after that. The residual effect of S is small unless it is applied as elemental S.

5.4.3   **Calcium, Magnesium, and Potassium**

Ca is mainly a "structural" element in cell walls and does not recycle in the vine. Mg is required mainly for chlorophyll and as an enzyme cofactor. K occurs mainly in ionic form and is the most mobile of the three elements. Fifty percent or more of the vine K can be redistributed to the fruit by harvest time. Sucrose accumulation and K transport appear to be coupled in the ripening berries. High K concentration in the berries is associated with high juice pH, which in turn leads to high wine pH. The influence of berry K concentration on wine quality is discussed in section 9.8.

About 2.5–4, 0.5–0.6, and 0.1–0.2 kg of K, Ca, and Mg, respectively, is removed per tonne of fruit. The higher K values occur in the hot Sunraysia and Riverland regions of southeastern Australia, where the deep sandy soils are relatively high in K.

Ca, Mg, and K are released by the weathering of feldspars, micas, and secondary minerals such as calcite, magnesite, and the micaceous clays. The presence of these minerals depends on the soil's parent material and its age. For example, Ca is abundant in soils formed on limestone and chalk, Mg is high in "greensands" (high in glauconite) and serpentinite, and K is high in soils formed on mica schists. Soils formed on basic igneous rocks (basalts, dolerites, and gabbros) are usually high in Ca and Mg. The pool of exchangeable Ca^{2+}, Mg^{2+}, and K^+ provides the immediate source of these elements for plants (section 4.6.1). Interlayer K^+ in micaceous clays is a more slowly available source. Total reserves are therefore variable, reflecting the conditions of soil formation, but exchangeable Ca^{2+} content is usually 1,000–5,000 kg/ha, Mg^{2+} is 500–2,000 kg/ha, and K^+ is ca. 1,000 kg/ha.

False K Deficiency. This disorder can occur on young vines in their third or fourth season, and it often coincides with periods of fluctuating temperatures before flowering. The lower leaves lose color, and the margins curl and become necrotic. The leaf K:N ratio, which is especially important in vines, is temporarily out of balance. Leaves with relatively low K concentrations frequently have high concentrations of the diamine putrescine, formed by breakdown of the amino acid arginine (and release of NH_4^+ ions).

Forms of K, Ca, and Mg Fertilizers. The commonly available K, Ca, and Mg fertilizers for vineyards are listed in table 5.8. KCl is the cheapest form, per unit of K, but K_2SO_4 is generally favored, especially in soils where salinity and high Cl are a problem. KNO_3 is the most expensive of the K fertilizers and should be used only where N as well as K is required. As with N, K must be controlled within well-defined limits to achieve good wine quality (section 9.8).

Losses and Residual Effects. There are no gaseous losses of these elements. Apart from crop removal, the main loss is by leaching; for example, Ca^{2+} is the main cation accompanying NO_3^- when it is leached from the soil. Many Australian duplex soils tend to have high exchangeable Mg^{2+} contents in the subsoil, which may lead to a K:Mg imbalance in deep-rooted vines. Conversely, a K-induced Mg deficiency may occur in high K soils with a K-accumulating rootstock, such as Ramsey.

Table 5.8   ***Common K, Ca, and Mg Fertilizers Used in Vineyards***

Compound	Formula	Element Content (%)	Comments
K			
Potassium sulfate	K_2SO_4	41	Easier to handle than KCl; more expensive per kg K than KCl, but preferred on saline soils
Potassium nitrate	KNO_3	38	Very soluble, supplies K and N
Muriate of potash	KCl	ca. 50	Very soluble in water; natural source is from salt deposits
Ca[a]			
Gypsum	$CaSO_4.2H_2O$	18–22	Sparingly soluble in water; primarily used to improve soil structure; contained in SSP
Mg[a]			
Epsom salts	$MgSO_4$	10	A water-soluble form of Mg fertilizer

[a]Ca and Mg are also supplied in liming materials (see table 5.9).

The residual effect of insoluble $CaCO_3$ and $MgCO_3$ is high. The residual effect of the more soluble forms depends on the *CEC* of the soil, and in the case of K, on the presence of mica clays that can fix K^+ in their structures (section 4.6.2).

5.5   *Soil pH and Nutrient Availability*

5.5.1   *The Macronutrients*

Soil pH influences the ratio of NH_4^+ and NO_3^- in the mineral N pool through its effect on nitrification. The optimum pH for nitrification in soil is 6.6–6.8, and the rate of nitrification is slow below pH 5. Above pH 7, the rate of oxidation of NO_2^- to NO_3^- decreases and NO_2^- accumulates—a condition that is undesirable for plant growth.

The effect of pH on soil P availability is discussed in section 5.4.2.1. Soil pH influences Ca, Mg, and K availability indirectly through its effect on mineral weathering, carbonate dissolution, and cation exchange. As the soil pH falls below 5, the concentration of exchangeable Al^{3+} increases and Al^{3+} ions displace exchangeable Ca^{2+}, Mg^{2+}, and K^+, which are then lost through leaching. Sulfate is also indirectly affected by soil pH. Positive charges on variably charged surfaces increase at low pH, so more sulfate is adsorbed and leaching is decreased.

5.5.2   *The Micronutrients*

Micronutrient availability is very dependent on soil pH. Over the pH range of 4–9, the elements Fe, Mn, Cu, Zn, and Co occur as cations. Mo occurs as an anion, and B as the very weak boric acid (H_3BO_3), which dissociates at pH >8 as follows:

$$H_3BO_3 + H_2O \leftrightarrow H^+ + B(OH)_4^- \tag{5.9}$$

As discussed in chapter 4, the availability of these elements is determined by the strength of their adsorption by clays and sesquioxides, relative to their ability to form complexes with organic compounds, and by their tendency to precipitate as insoluble oxides and hydroxides. The range of possible reactions is illustrated in figure 5.8. For vines, the most important micronutrients are Fe, for chlorophyll function, and Zn and B, for good fruit set.

The concentration of free micronutrient ions in the soil solution is normally very low. Although the surface adsorbed component (fig. 5.8) accounts for $<10\%$ of the total of any micronutrient in the soil, it is important because it buffers the soil solution against depletion. The cations Fe^{3+}, Mn^{2+}, Cu^{2+}, and Zn^{2+} are adsorbed on negatively charged surfaces in competition with Ca^{2+}, Mg^{2+}, and K^+ and also through complex formation on sesquioxide surfaces. As the pH increases, the micronutrient cations hydrolyze, which increases their affinity for the charged surface. The stronger retention contributes to the decrease in Fe^{3+}, Mn^{2+}, Cu^{2+}, and Zn^{2+} availability as the pH rises. However, hydrolysis of the metal cations is the first step in precipitation of the insoluble hydroxides, as discussed in appendix 5, thus Fe, Mn, Cu, and Zn deficiencies may occur in vineyards at a soil pH >7.

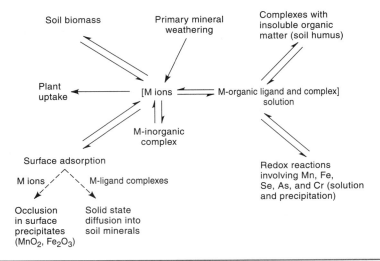

Surface adsorption

Figure 5.8   Possible reactions involving a micronutrient in the soil (White 1997). Reproduced with permission of Blackwell Science Ltd.

Conversely, MoO_4^{2-} is most strongly adsorbed at low pH on sesquioxides by a ligand-exchange mechanism analogous to that for $H_2PO_4^-$ (section 4.5.4). B is adsorbed the same way, as $B(OH)_4^-$. Because this ion only occurs at pH >8, B becomes less available at high pH levels.

Lime-induced Fe chlorosis is a common problem in vines growing on soils formed on chalk or limestone. The visual symptoms of chlorosis were introduced in section 5.2.1. French scientists have found that the total $CaCO_3$ content of the soil is not as important as the "active" $CaCO_3$, which is in a finely divided form (0–0.02 mm) (Baize 1993). The active $CaCO_3$ content is measured by extraction in ammonium oxalate and can be related to the Fe extracted by EDTA to give a *chlorosis-risk index (CRI)*, where

$$CRI = \frac{Active\ CaCO_3\ (\%)}{(Extractable\ Fe)^2\ (ppm)} \times 10^4 \tag{5.10}$$

Most of the American species of *Vitis* that provide phylloxera- and nematode-resistant rootstocks are more sensitive to lime-induced chlorosis than *V. vinifera*. Of the common rootstocks, only those derived from *V. berlandieri* show low sensitivity to chlorosis. Fe chlorosis can be treated by adding Fe-chelates to the soil (see appendix 5), but more commonly Fe, Mn, and Zn deficiencies are treated by spraying vines with soluble sulfates. Cu is applied in the fungicidal spray Bordeaux Mixture, which contains lime and $CuSO_4$. In older vineyards, Cu may accumulate to potentially toxic levels, especially in acid soils, because of the repeated spraying and strong complexing of Cu by soil organic matter.

Because of precipitation in insoluble compounds and strong sorption on surfaces, *micronutrient losses from leaching* are very small. The exceptions are B, which is weakly adsorbed up to pH 8, and Fe and Mn, which can be leached in their reduced forms in strongly gleyed soils (section 1.3.3.2). Nevertheless, because atmospheric inputs are small (except in heavily industrialized areas), cumulative losses over time from very old soils that are sandy may lead to micronutrient deficiencies, as in some of the coastal regions of southern and southwestern Australia.

5.5.3   *Correction of Soil Acidity*

Soil acidification is a natural process, but it can be accelerated under viticulture, especially when NH_4-N fertilizers and S-based fungicides are used. If concentrations of exchangeable Al^{3+} in acid soils exceed 1 mg/kg soil, root growth is inhibited.

To correct soil acidity, various liming materials based on $CaCO_3$ are used, depending on their availability and price (table 5.9). The dissolution of $CaCO_3$ in soil is an alkali-producing reaction, according to the equation

$$CaCO_3 + CO_2 + H_2O \leftrightarrow Ca(HCO_3)_2 \qquad (5.11)$$

which shows that dissolution depends on the partial pressure of CO_2 in the soil air. At a partial pressure of 0.0365 kPa (atmospheric) and a Ca concentration in solution of 1 mmol/L, the equilibrium pH attained in the presence of solid $CaCO_3$ is 8.4.

The amount of lime required to raise the pH to a desired value depends on the soil's titratable or total acidity (section 4.6.3.3). This amount is called the soil's *lime requirement*, the measurement of which is discussed in box 5.6. Lime is best applied during autumn so that it begins to dissolve in the moist soil during winter while the vine is dormant.

Table 5.9   ***Liming Materials and Their Neutralizing Value***

Material and Chemical Composition	Neutralizing Value (NV)[a] (%)	NV of Commercial Grade (%)	Comment
Burnt lime, CaO	179	<120	Reacts vigorously with water
Hydrated lime, Ca(OH)$_2$	135	<105	Occurs as a very fine powder; difficult to handle
Dolomite, (Ca,Mg)CO$_3$	109	100–119	More soluble than calcium carbonate, ca. 11% Mg
Limestone, CaCO$_3$ plus impurities	—	50–85	NV depends on concentration of impurities such as clay, silica, sesquioxides
Kiln dust, CaCO$_3$, CaO, some K	—	90–110	By-product of cement manufacture; fine powder, often pelletized

[a]Calculated on the basis of the pure material, relative to calcite, which is 100% $CaCO_3$
Source: Data from Goldspink (1996) and Cass (1998)

| Box 5.6 | *Measuring a Soil's Lime Requirement* |

The *lime requirement* is the quantity of ground limestone or chalk (t/ha to a depth of 15 cm) required to raise the soil pH to a desired value. In Australia, a pH of 6.5 (1:5 in H_2O) is recommended for vineyards (Goldspink 1996). However, in parts of Beaujolais and Languedoc-Rousillon in France, grapevines have been grown successfully on acid soils (as low as pH 4) for many years. At minimum, sufficient lime should be applied to hydrolyze the exchangeable Al^{3+} (to <1 mg/kg), which is normally achieved at pH 5.5. Gypsum does not have a liming effect, but the Ca^{2+} supplied competes with Al^{3+} during ion uptake and hence ameliorates the adverse effects of Al.

Methods of measuring the lime requirement range from equilibrating a soil sample with a single buffer solution to titrating the soil with $Ca(OH)_2$. Details of laboratory methods are given in Rayment and Higginson (1992). Soils with a small pH buffering capacity, such as sandy loams poor in organic matter, have a smaller lime requirement than soils with a high pH buffering capacity, such as clays and organic-rich loams. The lime requirement ranges from 1–2 t $CaCO_3$/ha for sandy loams to 4–5 t $CaCO_3$/ha for clay soils.

Laboratory analysis is also necessary to measure the *neutralizing value* (NV) of the liming material. The standard in Australia is pure $CaCO_3$. The NV of various liming materials is given in table 5.9. Another important property is the fineness of the liming material, because the smaller the particle size, the faster the reaction with the soil. A high-quality liming material should have an NV > 75% with 80% of the particles <0.6 mm diameter. To be most effective, lime should be cultivated into the surface soil.

5.6   *Reactions Affecting Nutrient Availability in Waterlogged Soil*

5.6.1   *Anaerobic Processes*

When the soil pore space fills with water, the soil becomes waterlogged. Soil aeration is then inadequate to sustain aerobic respiration (section 3.4), and microorganisms capable of respiring anaerobically begin to multiply. Carbohydrates are fermented to organic acids such as acetic and butyric acids, which are subsequently metabolized to the gases CO_2 and CH_4, with some ethylene (C_2H_4) and H_2 also produced. Fermentation releases much less energy for microbial growth than aerobic respiration. Although low concentrations of C_2H_4 may stimulate root elongation, concentrations >1 mg/L in the gas phase may inhibit root growth.

In addition to their direct effect on root metabolism and growth, anaerobic conditions predispose to a chain of reactions that can have undesirable effects. These are called *redox reactions* because they involve the transfer of electrons (e^-) produced by respiration to acceptors other than O_2. The first of these reactions—NO_3^- reduction, otherwise called *denitrification*—can occur in partially waterlogged soil, that is, when the macropores between aggregates still contain O_2, but the small pores inside the aggregates are anoxic (see fig. 3.7). The complete reaction for denitrification is

$$2NO_3^- + 12H^+ + 10e^- \leftrightarrow N_2 \text{ (gas)} + 6H_2O \qquad (5.12)$$

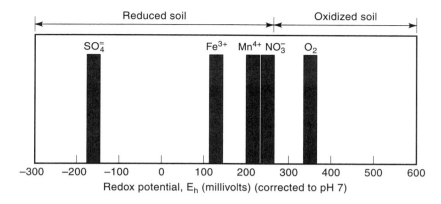

Figure 5.9 Threshold redox potentials at which the oxidized forms shown become unstable (White 1997). Reproduced with permission of Blackwell Science Ltd.

The gas N_2O is an intermediate product, and at pH <5, may be comparable in concentration to N_2. But at pH >6 and temperature >25°C, most of the N_2O is reduced to N_2, especially when the soil approaches complete saturation. Denitrification is carried out by facultative anaerobic bacteria, such as *Bacillus* and *Pseudomonas* species.

As the NO_3^- is depleted, reducing conditions intensify, and other compounds become susceptible to reduction in the following sequence:

$$MnO_2 \text{ (solid)} + 4H^+ + 2e^- \leftrightarrow Mn^{2+} + 2H_2O \tag{5.13}$$

$$Fe(OH)_3 \text{ (solid)} + 3H^+ + e^- \leftrightarrow Fe^{2+} + 3H_2O \tag{5.14}$$

$$SO_4^{2-} + 10H^+ + 8e^- \leftrightarrow H_2S \text{ (gas)} + 4H_2O \tag{5.15}$$

The intensity of reducing conditions, or the reducing power, is measured by the *redox potential E_h*, which is explained in appendix 6. The approximate E_h value at which the oxidized form of each substance (on the left-hand side of the equations) becomes unstable is shown in figure 5.9. The soil is *poised* in a particular E_h range until most of the oxidized form has been consumed, and then the soil E_h drops to a lower value as the reducing power intensifies. Reactions such as 5.12 and 5.15, in which a gas of low water solubility is produced, are irreversible. The critical E_h for damage to vine roots from anaerobic conditions is $\leq$ 100mV.

5.6.2 Redox Reactions and Nutrient Availability

Denitrification is a loss of available N (as NO_3^-). Any N_2O produced during denitrification is undesirable because it is a potent "greenhouse gas." Because hydrated iron oxides are plentiful, many waterlogged soils are poised by the Fe^{3+}-Fe^{2+} redox system. The change from the red-brown color of Fe_2O_3 to the blue-grey color of Fe^{2+} forms is typical of a gleyed soil. Both Mn and Fe are more soluble in the reduced form and may be leached. The dissolution of MnO_2 can also release coprecipitated Co into the soil solution. Further, the solubility of elements such as Co, Ni, Mo, and Cu (which are not reduced at the E_h values attained in soil) may be increased under waterlogged conditions, because of the faster weathering of ferromagnesian minerals.

Note that H^+ ions are consumed in reducing reactions, so the soil pH tends to rise. This can affect the availability of micronutrients, and also P. For example, in waterlogged soil, reduction of Fe^{3+} in iron oxides releases adsorbed P or P from insoluble $FePO_4$. However, at pH >6, Fe^{2+} compounds become less soluble, and a metastable iron oxyhydroxide, $Fe_3(OH)_8$, precipitates. This precipitate provides a highly reactive surface for the readsorption of P, and certainly all the dissolved P is readsorbed when aerobic conditions return and $Fe(OH)_3$ precipitates.

At very low E_h values (fig. 5.9), SO_4^{2-} is reduced due to the activity of obligate anaerobic bacteria of the genus *Desulphovibrio*. This reaction depletes the available sulfate supply. As the concentration of H_2S increases (reaction 5.15), ferrous sulfide (FeS) precipitates and slowly reverts to *iron pyrite* (FeS_2). Iron pyrite is a mineral found in present-day estuarine and marine sediments. Over geological time, such sediments have transformed into sedimentary rocks, often with associated coal deposits. For example, FeS_2 occurs in the *cendres noires* that have been regularly applied to build up the Chalk soils of the Champagne region in France (section 1.3.3.2). When exposed to air, FeS_2 oxidizes according to the reaction

$$FeS_2 + 15/4O_2 + 7/2O_2 \leftrightarrow Fe(OH)_3 + 4H^+ + 2SO_4^{2-} \qquad (5.16)$$

In chalky and calcareous soils, the H_2SO_4 formed reacts with $CaCO_3$ to create microsites of lower pH in the mineral soil, where Fe is more available.

5.7   *Organic Viticulture*

Organic viticulture involves growing grapes without the use of industrially manufactured fertilizers (sometimes called "chemical fertilizers") and herbicides, pesticides, or fungicides for weed, pest, and disease control. France is probably the home of organic viticulture and "organic wine," but the practice is becoming more popular in many of the world's wine regions. This process has also been referred to as "sustainable viticulture" or "alternative viticulture." An ideological variant of organic viticulture is "biodynamic viticulture," which follows the same basic tenets as organic viticulture, but in addition holds that the moon and stars influence the vine and growth of its parts at different times of the day and year. The scientific basis of these beliefs has not been established.

Strictly, a vineyard can be called "organic" only if it has been registered, inspected, and certified by a supervisory organization such as Ecocert, Nature et Progrès, or Terre et Vie in France, the Organic Foods Production Association of North America, or the Organic Vignerons Association of Australia. Normally, a three-year conversion period is required for vineyards to be certified as organic. Certified organic production means that grapes are grown without insecticides, herbicides, fungicides, and chemical fertilizers, other than those approved by the certifying agency. Nor can genetically modified vines be grown or genetically modified yeast used in the fermentation to produce organic wine.

Many vignerons have not adopted true organic certification, which is an exacting standard, but have moved toward minimum intervention in the vineyard and using natural materials where possible. This approach involves minimizing pesticide and fungicide sprays in a system of integrated pest management (IPM), which encourages a balance between natural predators and pests, and good canopy management to avoid disease problems. It is easier to avoid fungal attack in warm, dry climates such as the south of France than in cooler, wetter regions such as Burgundy or Alsace-Lorraine. Organic manures, composts, and mulches are used instead of chemical fertilizers, with the dual purpose of supplying nutrients and improving the biological health of the soil. Biodiversity is increased through the use of cover crops that include a range of species and help to improve soil structure (section 7.3.1).

If the cover crop contains legume species, its decomposition provides mineral N to the vines. Plowing-in also removes competition between the vines and cover crop for soil water, which is important if an organic grower prefers not to irrigate. Generally, organic viticultural practices reduce the adverse impacts of grape growing on the environment.

Organic Manures and Soil Amendments. The excreta of farm animals, composted plant material, and biosolids (sewage sludge) are classed as *organic manures*. Biosolids are generally prohibited for use in organic viticulture, unless specifically exempted by the certifying authority.

Farm Animal Manure. The most commonly used manure in vineyards is poultry manure, which is plentiful and easily collected because the birds are kept in confined spaces. Fresh poultry manure (from caged birds) is high in N, mainly as uric acid, which hydrolyzes to urea and subsequently to NH_3 gas. Large losses of N can result when the manure is stacked in heaps, and after it is spread on the soil. Poultry manure produced from deep litter systems (where the droppings are mixed with straw or sawdust) is lower in N and has a higher C:N ratio. Thus, much of the N becomes immobilized by microorganisms. As decomposition proceeds in the soil, the C:N ratio falls and N will eventually be remineralized. The range of N, P, K, and micronutrient content in poultry manure is shown in table 5.10.

Compost. Compost is made by accelerating the decomposition of plant residues in well aerated, moist heaps. Composted green garden waste from cities is increasingly being used in vineyards. Compost can also be made from the "pomace" (pressed skins from the winery, called "marc" in Australia) or from nonin-

Table 5.10　　*Composition of Some Manures, Organic Fertilizers, and Soil Amendments Suitable for Organic Viticulture*

Material	N	P	K	Ca	Mg	Cu	Zn	Mn
	(%)[a]					(mg/kg)[a]		
Pelleted poultry manure	3–6	1–2	1.3–1.5	2–6	0.4–0.6	30–170	220–290	300–370
Compost	2.1–3.9	0.6–1.2	0.6–2.7					
Blood and bone	7–8	4–5		13–14				
Ground rock dust				5	1–2.2	50–70	10–150	600–800

[a]Expressed on a DM basis. Nutrient contents are variable and should be confirmed by analysis.
Source: Data from Goldspink (1996) and White (1997)

Figure 5.10   Composting of garden waste in windrows. Photograph by the author. See color insert.

dustrial sewage sludge, provided bulking-up material such as wood chips or sawdust is added. The ideal C:N ratio for successful composting of such mixtures is about 20. The composting is done in heaps or windrows (fig. 5.10), where the temperature should reach at least 55°C for 3 days to kill weed seeds and pathogens. Well-made compost should be odorless. Organic manures and compost are best spread along the rows in winter when the vines are dormant. Compost can serve as a mulch that reduces in-row soil temperatures and evaporation rates—an important function in regions with hot summers.

"Compost tea" is an organic-rich liquid that drains from a compost heap. Although it is sometimes applied as a foliar spray in organic vineyards, its benefits are unproven.

Organic Fertilizers and Soil Amendments. By-products from the animal processing industry may be used under regulated conditions in organic viticulture. The example of blood and bone is given in table 5.10. RPRs are an acceptable source of P, but the conditions controlling their effectiveness should be observed (section 5.4.2.1). For example, mixing RPR with lime is neither necessary nor desirable because an increase in pH slows the rate of dissolution of the RPR, and dissolution of the RPR itself has a liming effect. Further, because RPR is insoluble in water, when used as a foliar spray, it is ineffective in supplying P to the vines.

Cu supplied in Bordeaux Mixture for the control of downy mildew is acceptable in organic viticulture. Elemental S, which is used as a wettable powder in sprays to control powdery mildew, is an acceptable source of S. Wood ash is a suitable source of K (as is K_2SO_4, but not KCl). Ground rock dust can supply Ca, Mg, and some of the micronutrients (table 5.10), but should be incorporated into the soil. Proprietary brands of seaweed extract may also be used. Although the amount of nutrients supplied at normal application rates is small, the vines (and soil microorganisms) may benefit from growth-regulating compounds in the extracts.

5.8 *Summary Points*

This chapter on nutrients and their supply in fertilizers is summarized as follows:

▮ Vineyard productivity depends on soil fertility, which is determined by the nutrient-supplying power of the soil. Although visual symptoms in vines can be observed in the field, deficiency or excess of macro- and micronutrients in grapevines is usually diagnosed by *plant analysis* or *tissue testing*. The *critical value* is the nutrient concentration at which the vine no longer responds to further additions of that nutrient.

▮ A basal leaf, opposite a bunch near the base of a shoot, is sampled on a vine at full flowering. Some 75–100 leaves from vines of one variety in a block should be collected to make a *representative sample* for analysis. Petioles are preferred to leaf blades. A further sample can be collected at veraison to check on doubtful results. Samples should not be taken soon after any sprays or dusts have been applied.

▮ When a vineyard is being established, soil analyses, especially for P, micronutrients and pH, must be used to assess nutrient requirements. Soil samples (to 50-cm depth) can also be analyzed to measure the amount of mineral N (NH_4^+ and NO_3^-) in the root zone. However, soil N supply during the growing season is mainly determined by net mineralization of organic N. *Green manuring* with leguminous cover crops produces a short-term increase in soil organic N. The rate of N mineralization can be estimated from the organic N content and decay coefficients.

▮ Both plant and soil tests must be calibrated against the grape yield response to a particular fertilizer. From this, the quantity of nutrient required is assessed. By comparing the value of the product with the cost of individual fertilizers, the most appropriate form and optimum quantity of fertilizer can be determined.

▮ Yields and soil properties often vary widely within one vineyard. This variation can be measured annually and mapped in a GIS, which enables specific management to be applied within blocks, or between blocks, to remove yield and/or grape quality constraints. This is called *precision viticulture*.

▮ The amount and timing of N fertilizers (mainly $Ca(NO_3)_2$, urea, and NH_4NO_3) must be carefully matched to the vines' needs and to the period of most rapid root growth. Too much N leads to excess vigor and the risk of surplus NO_3^- being leached below the root zone. Too little N leads to reduced yields and to low *yeast assimilable N (YAN)* in the fruit, which in turn can lead to "stuck fermentations." The optimum *YAN* in grape juice is between 200 and 300 mg N/L.

▮ Soluble N fertilizers can be applied in irrigation water (*fertigation*), which is the best way to avoid volatilization of NH_3 from urea spread on the soil surface, especially for calcareous soils.

▮ P can be supplied as *water-soluble fertilizers* (SSP, TSP, MAP, and DAP) or as insoluble *reactive phosphate rocks* (RPRs). To dissolve, RPRs depend on the soil to supply H^+ ions and moisture. As they dissolve, they have a small liming effect. However, the main liming materials are ground limestone and chalk ($CaCO_3$), which should be cultivated into the soil to accelerate their effect in raising soil pH. The amount of lime needed to raise the soil pH to the desired value is called the *lime requirement*. The desired pH is in the range 5.5–6.5, high enough to eliminate any inhibitory effect of Al^{3+} ions on root growth and to optimize micronutrient availability.

▪ S is supplied in SSP, gypsum, or K_2SO_4, and also as elemental S in wettable dusts for powdery mildew control. N and S can also come from the atmosphere in rain or as dry deposition. P has a long residual effect in soil as a result of *P fixation*, and does not leach. S leaches as SO_4^{2-}, and does not have a long residual effect (unless it remains in the elemental form).

▪ K is supplied as K_2SO_4, KNO_3, and KCl, though extra input of Cl should be avoided in soils with some salinity. High concentrations of tissue K ($>2\%$ DM) must be avoided because of adverse effects on berry pH and wine quality.

▪ The availability of micronutrients is influenced by soil pH. The cations Fe^{3+}, Mn^{2+}, Cu^{2+}, and Zn^{2+} are more available at low pH, whereas the anion MoO_4^{2-} is more available at high pH. B exists as H_3BO_3, and as $B(OH)_4^-$ at high pH. Micronutrient deficiencies are most commonly corrected with foliar sprays. Lime-induced Fe chlorosis on calcareous soils depends on the fraction of "active" (finely divided) $CaCO_3$.

▪ Waterlogging should be avoided, not only because of its direct impact on vine growth but also because of undesirable redox reactions, such as denitrification, reduction of Mn and its increased solubility (to potentially toxic levels), and reduction of SO_4^{2-}.

▪ Organic viticulture involves growing grapes without the use of manufactured chemicals as fertilizers or for weed, pest, and disease control. Certified organic vineyards must be registered and inspected by recognized organizations. But many vignerons are voluntarily reducing their reliance on pesticides and fungicides through *integrated pest management*.

▪ Organic manures (mainly poultry manure), composted plant material, and cover crops can potentially be used in organic viticulture. RPRs, elemental S, and Bordeaux Mixture (a traditional Cu spray) are also acceptable.

6 Soil–Water–Vine Relationships and Water Management

6.1 Water Quantity and Water Energy in Soil

Water is a prerequisite for vine growth. It is essential for photosynthesis and to maintain the hydrated conditions and cell turgor necessary for a host of other biochemical processes in the plant. As we saw in chapter 4, diffusion of nutrient ions to the root, and their movement by mass flow into the vine's "transpiration stream," both depend on water.

The *volumetric water content* θ, defined as the volume of water per unit volume of soil (section 3.3.2), indicates how much water the soil can hold. However, to understand what drives water movement in the soil, we must understand the forces acting on the water because they affect its *potential energy*. The energy status of soil water also influences its availability to plants. There is no absolute scale of potential energy. But we can measure changes in potential energy when useful work is done on a measured quantity of water or when the water itself does useful work. These changes are observed as changes in the *free energy* of water, which gives rise to the concept of *soil water potential*. The derivation of the soil water potential ψ (psi) is given in appendix 7.

Historically, the energy status of soil water has been described by a number of terms related to soil water potential, such as *pressure*, *suction*, or *hydraulic head*. These terms and their units are explained in box 6.1. The terms ψ and *head* will be used in this book.

6.1.2 Components of the Soil Water Potential

Several forces act on soil water to decrease its free energy and give rise to *component potentials*. These are adsorption forces, capillary forces, osmotic forces, and gravity.

141

Box 6.1   *Terms Describing the Energy Status of Soil Water*

Soil water potential ψ, as defined in appendix 7, is expressed as energy per mole (Joules (J)/mol). To be compatible with older pressure measurements, ψ is converted to energy per unit volume (J/m³) by dividing by V_w, where V_w is the volume (m³) per mole of water. The units of J/m³ are equivalent to newtons/m² or pascals (Pa), which are units of *pressure*. In saturated nonsaline soil, ψ is zero.

As water is withdrawn from a wet soil, ψ becomes negative. In this case, ψ can be thought of as a negative pressure. A negative pressure implies a suction force acting on water to draw it into dry soil. Thus, soil water potentials are often referred to as *suctions*, when the negative sign is dropped. The term suction is meaningless when ψ is >0 (for example, in saturated soil below a watertable). In this situation, the soil water pressure is positive. The normal range for ψ in soil is from 0 to -1500 kPa.

To describe water movement in soil, hydrologists and soil scientists usually express the water potential in terms of *hydraulic head*, or simply *head h*. Then ψ is expressed as energy per unit weight of water (J/kg). Head h is defined by the equation

$$h = \frac{\psi}{\rho g} \tag{B6.1.1}$$

where ρ = the density of water (1000 kg/m³) and g = the acceleration due to gravity (9.8 m/s²). If the soil water potential is 10^5 Pa (= 1 bar, the normal pressure of the atmosphere), the equivalent head is given by

$$h = \frac{100,000}{1000 \times 9.8} = 10.2 \text{ m} \tag{B6.1.2}$$

Head units are the same as units of length (m). Hydraulic head expressed in cm normally ranges from 0 to $>15,000$ cm. To make these numbers less unwieldy, the *pF scale* of soil water potential was introduced. The pF value is defined as $\log_{10}h$ (in cm). The pF scale is not used much at present.

Comparative values for the energy status of soil water in potential, head, and pF units at several states of wetness are given in table B6.1.1.

Table B6.1.1   *Soil Water Potential, Hydraulic Head, and Soil pF at Different Wetness States*

State of Soil Wetness	Water Potential ψ (kPa)	Head h (m)	pF Value
Saturated, nonsaline soil	0	0	—[a]
Water held after free drainage for 48 hours (field capacity)	-10	1.02	2.0
Approximate water content at which plants wilt (wilting point)	$-1,500$	153	4.2
Soil at equilibrium with a relative vapor pressure of 0.85 (approaching air-dryness)	$-22,000$	2244	5.4

[a]Undefined

Adsorption Forces. In very dry soils (relative humidity, RH, of the soil air <20%), water is adsorbed onto the clay and silt particles as a monolayer in which the molecules are hydrogen bonded to each other and the surface. With an increase in RH, more water molecules are adsorbed by hydrogen bonding to those on the surface. The charged surfaces of clay minerals also attract cations, and the electric field of the cation orients the polar water molecules around the ion to form a hydration shell, containing 6–12 water molecules. Work is done in attracting water molecules to a cation, which means there is a reduction in the free energy of water in each hydration shell.

Capillary Forces. As water layers build up in a dry soil, water is drawn by surface tension forces into the narrowest pores and wedges between soil particles, forming curved air–water menisci (fig. 6.1a). The effect on the soil water is exactly analogous to the rise of water in a glass capillary tube inserted into a dish of water (fig. 6.1b). The decrease in free energy as the surface tension force draws water up the tube is balanced by the increased potential energy of the water column. Atmospheric pressure acts on the upper surface of the water in the capillary tube and also on the free water surface outside the tube. Thus, the pressure difference ΔP across the meniscus is equal to the decrease in free energy of the wa-

Figure 6.1   (a) Water held by surface tension forces in narrow pores between soil particles (b) Water drawn up a glass capillary tube by surface tension forces (White 1997). Reproduced with permission of Blackwell Science Ltd.

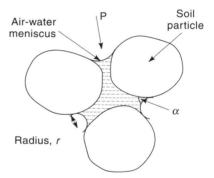

(a)

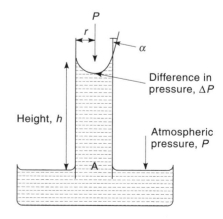

(b)

ter ψ (= $h\rho g$ from equation B6.1.1). When the water is adsorbed to a very clean glass surface, the angle of wetting α (alpha) is assumed to be zero, so we have

$$\Delta P = h\rho g = \frac{2\gamma}{r} \tag{6.1}$$

where h = the height of the water column, r = the radius of the capillary tube, and γ (gamma) = the surface tension of water (force per unit length). Not all soils can be assumed to have a zero angle of wetting: where α is significantly greater than 0, we have nonwetting soils, as explained in box 6.2.

Summary. The combined effect of adsorption and capillarity on the free energy of soil water is expressed through the *matric potential, ψ_m.* Over the normal range of potentials in nonswelling soils (0 to -1500 kPa), the matric potential ψ_m can be related directly to the pore radius through equation 6.1. Table 6.1 shows the radius of the largest pores that would just hold water at ψ_m values corresponding to the soil wetness states in table B6.1.1.

Osmotic Forces. Dissolved solutes (ions and molecules) also reduce the free energy of water through the action of an osmotic force. The reduction in free energy is measured by the *osmotic potential, ψ_s.* Osmotic potential is numerically equal (but opposite in sign) to the osmotic pressure of the soil solution. The os-

---

Box 6.2   *Nonrepellent and Water-repellent Surfaces in Soil*

The angle of wetting or *contact angle* α is measured from the liquid–solid interface to the liquid–air interface. At most soil mineral surfaces, the angle is close to 0 and the water is said to "wet" the surface (fig. B6.2.1a). However, if these surfaces become coated with hydrophobic organic compounds, the water molecules are more strongly attracted to each other than to the mineral surfaces. The water forms distinct droplets and the contact angle increases markedly (fig. B6.2.1b). The surface is said to "repel" the water, and the whole soil may become *water repellent.* Such an effect occurs in sandy surface soils that experience severe drying. Water subsequently applied to the surface neither wets the surface, nor infiltrates— rather, water runs off. Many soil surfaces may exhibit some degree of water repellency if they remain air-dry for a prolonged period, but the effect gradually disappears as the soil wets up again.

Figure B6.2.1 (a) An air-water-solid interface with a small contact angle: water "wets" the surface. (b) An air-water-solid interface with a large contact angle: water is "repelled" by the surface (White 1997). Reproduced with permission of Blackwell Science Ltd.

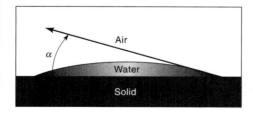 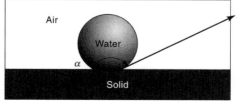

Table 6.1   **Maximum Radius of Cylindrical Pores Holding Water as a Soil Dries from Saturation to Air-dryness**

State of Soil Wetness	Soil Water Potential ψ (kPa)	Pore Radius (μm)
Saturation	0	infinitely large
Water held after free drainage for 48 hours (field capacity)	−10	15
Approximate water content at which plants wilt (wilting point)	−1,500	0.1
Soil at equilibrium with a relative vapor pressure of 0.85 (approaching air-dryness)	−22,000	0.007

motic pressure π (pi) is defined as the hydrostatic pressure necessary to just stop the inflow of water when the solution is separated from pure water by a semipermeable membrane. Osmotic pressure is calculated from the equation

$$\pi = RTC \tag{6.2}$$

where R and T are defined in appendix 7, and C is the concentration of solute particles in solution. When C has the units mol/m^3, π is in pascals. The component ψ_s is significant only in saline soils.

Gravity and Weight. Gravity exerts a force on water that is directly proportional to the height of the water above or below a chosen reference level. This gives rise to the component *gravitational potential* ψ_g. If the soil water is at a height above that chosen for the reference level, usually the soil surface, ψ_g is positive. In addition, soil water may experience a pressure greater than atmospheric pressure, the *pressure potential* ψ_p. In practice, variations in ψ_p due to changes in air pressure are negligible: ψ_p becomes significant only when part or all of the soil profile is saturated, as, for example, when groundwater rises into the soil. The upper surface of the groundwater is called the *water table*, and water below the water table experiences a hydrostatic pressure (equal to ψ_p) that is directly proportional to the depth h (see equation B6.1.1).

As stated previously, in saturated soil ψ_m must be zero, and it becomes negative as the soil dries (or the water table falls). Thus, ψ_m and ψ_p are two subcomponents of a pressure-matric potential continuum for which

- above the water table, $\psi_p = 0$ (at atmospheric pressure) and ψ_m is negative;
- at the water table, $\psi_m = \psi_p = 0$; and
- below the water table, $\psi_m = 0$ and ψ_p is positive.

Summary. The total water potential ψ in soil is the sum of the component potentials, that is,

$$\psi = \psi_m + \psi_s + \psi_p + \psi_g \tag{6.3}$$

In nonsaline soils, equation 6.3 simplifies to

$$\psi = \psi_m + \psi_p + \psi_g \tag{6.4}$$

Soil water potential is measured in the field using a tensiometer or a resistance block. These measurements are discussed in box 6.3.

Box 6.3 *Measurement of Soil Water Potential in the Field*

Tensiometers and resistance blocks measure the combined effect of ψ_m and ψ_g. Tensiometers operate in the range 0 to -85 kPa, and resistance blocks from -10 to -600 kPa. A tensiometer consists of a porous ceramic cup (with an air-entry value of at least 85 kPa) sealed to a length of polyvinyl chloride (PVC) tubing. The tube and cup are filled with water so that when the tensiometer is installed, the water in the tensiometer comes to equilibrium with water in the surrounding soil. The suction on the soil water (less the osmotic effect of any solutes that diffuse freely through the cup walls) is transmitted through the water column, to be read directly with a pressure gauge or transducer sealed to the top (fig. B6.3.1, left). The units of measurement are kPa or the equivalent centibars. Thus, the *tensiometer potential* ψ_t measures the soil water potential ψ minus any osmotic component. This is satisfactory for all but very saline soils.

Resistance blocks consist of two electrodes embedded in a porous material. Traditionally, this porous material was gypsum. But now synthetic materials more durable than gypsum are used. An example is the "Watermark" sensor that functions in the ψ range -10 to -200 kPa. These are also called "Lite" sensors for use in "light soils" (fig. B6.3.1, right). The more traditional sensor is a "heavy" sensor for use in "heavy soils." Water moves in or out of the block in response to a change in ψ in the surrounding soil, until equilibrium is established. The change in water content causes a change in electrical resistance of the block. The relationship between electrical resistance and water potential ($= \psi$) for each block must be determined for individual blocks—this is called calibration. Ions diffuse into and out of a block, which can affect its electrical resistance. Lite and heavy sensors are not affected by salinity up to 1 decisiemen (dS)/m and 6 dS/m, respectively (Goodwin 1995), so gypsum blocks can be used in quite saline soil. The measurement of soil salinity is discussed in chapter 7.

Figure B6.3.1 A tensiometer with vacuum gauge and a set of "Lite" gypsum blocks. Photograph by the author.

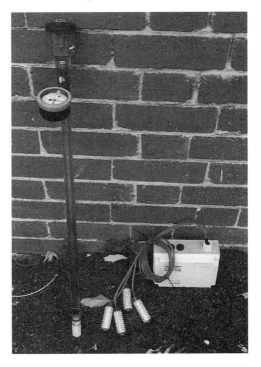

6.1.2   *Hydraulic Head of Water in the Field*

Equation 6.4 is often expressed in head units. Take, for example, the case where a pipe open at both ends is installed in the soil to a depth z below the soil surface. This is called a *piezometer*. If the soil is saturated at point A (fig. 6.2a), we may wish to know the water potential in head units at this point. Noting the continuum between pressure potential and matric potential, when the soil is saturated, we define the *piezometric head h* as

$$h = p - z \qquad (6.5)$$

where p is the positive pressure in the water at point A. The negative sign appears in this equation because, by convention, z is measured positively upward from the reference level.

Alternatively, a tensiometer installed in unsaturated soil at a depth z below the soil surface (point B in fig. 6.2b) will indicate that the water is at a negative pressure or *suction*. Water in the tensiometer equilibrates with the soil water surrounding the porous pot (at point B) such that the head h of the soil water is given by

$$h = -p - z \qquad (6.6)$$

Head gradients are used in calculating the rate of water movement through soil in the field (section 6.3). But before considering rates of water flow, we should examine the relationship between θ and ψ.

Figure 6.2   (a) A piezometer showing soil water at positive pressure in saturated soil. (b) A tensiometer measuring the negative pressure of soil water in unsaturated soil (redrawn from White 1997).

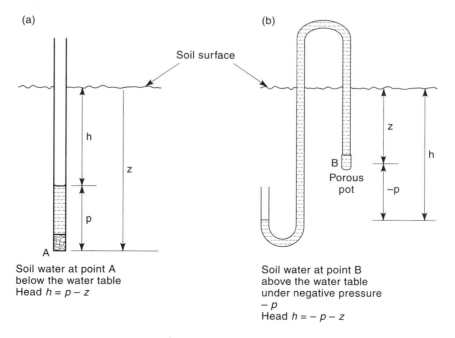

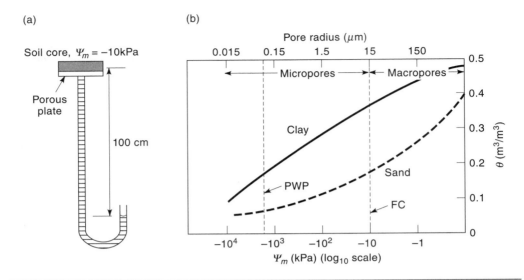

Figure 6.3   (a) Soil water under suction on a porous plate attached to a "hanging" water column. (b) Typical soil water retention curves for a clay soil and a sandy soil, showing the normal ψ_m range and corresponding pore radii.

6.2   *Water Content versus Water Potential Relationships*

6.2.1   *The Soil Water Retention Curve*

If a saturated soil core is placed on a porous plate attached to a hanging water column, θ falls because of the decrease in ψ_g created by the difference in height between the soil and the open arm of the water column (fig. 6.3a). The soil water now has a greater ψ_g than the open arm of the water column; this is compensated for by a decrease in ψ_m in the soil. This kind of apparatus can therefore be used to show how θ changes with a decrease in ψ_m. The practical limit of ψ_m is about -10 kPa (a head of -100 cm). For greater reductions in ψ_m, the soil is usually brought to equilibrium with an applied air pressure that is then equal but opposite to ψ_m in the soil pores. This is done using a pressure plate apparatus. A graph of θ against ψ_m called a *soil water retention* curve, can then be plotted (fig. 6.3b). The curves represent a sandy loam soil and a clay soil. These curves show that the sandy loam soil releases a larger proportion of its water at relatively high ψ_m values, whereas the clay soil releases water more gradually over the full range of ψ_m. This is primarily due to the difference in pore-size distribution between the two types of soil, as discussed in box 6.4.

The pore-size distribution influences not only water retention in soil, but also the rate at which water moves through soil (section 6.3).

6.2.2   *Available Water Capacity*

Although the soil water retention curve indicates a continuous relationship between θ and ψ_m, for practical purposes we recognize two fixed points in the rela-

---

Box 6.4   *Estimating the Pore-size Distribution of a Soil*

Equation 6.1 shows that the potential ψ (= $h\rho g$) is inversely proportional to the radius of the largest pores holding water at that potential. Using this relationship, the radius of pores $r - \delta r$ (delta r, where δr is a small change in r) that release water for a small decrease in potential, $\delta \psi_m$, is known. The volume of these pores can be calculated from the change in θ of the soil. Thus, the slope $d\theta/d\psi_m$ of the soil water retention curve, as shown in figure 6.3b, can be used to define the soil's pore-size distribution (provided that shrinkage of the soil as it dries is small). The larger the value of $d\theta/d\psi_m$, the greater the volume of pores holding water within the size range defined by δr. The curves in figure 6.3b show that the sandy loam soil has a majority of large pores (macropores) that release water at low suctions, compared with the clay soil, which has mainly small pores within aggregates (micropores) that release water at high suctions. Macropores promote drainage and aeration, whereas pores within aggregates are more likely to hold water in the "available range" (section 6.2.2).

---

tionship. The volume of water held between these two points defines the soil's *available water capacity, AWC*, which is water nominally available for plant growth. The upper limit of *AWC* is set by the field capacity *FC* (section 3.3.3). This point corresponds approximately to the water content at which the macropores are drained, but micropores within aggregates remain filled with water. Measurements with tensiometers show that soils attain *FC* at ψ_m values between -5 and -10 kPa. The value of -10 kPa is accepted for soils in Australia.

The lower limit of *AWC* is set by the θ value at which plants wilt, that is, the *permanent wilting point, PWP*. Wilting indicates that the rate of transpiration exceeds the rate of water movement to the roots, and the plant loses turgor. For many plants, the critical leaf water potential, ψ_l, at which wilting occurs is about -1500 kPa (-1.5 MPa). The onset of permanent wilting depends not only on ψ in the soil, but also on the plant's ability to lower ψ_l to maintain a favorable gradient for water flow, and hence rate of water flow, from the soil to the root. Thus, both *FC* and *PWP* depend to some extent on the dynamics of water movement in the soil, as discussed in section 6.3.

6.3   *Dynamics of Soil Water*

6.3.1   *Infiltration and Runoff*

Water entry into soil is called *infiltration*. The water arrives as rain or irrigation, but the vigneron has little control over the former. Whether all the rain that falls is absorbed depends on the condition of the soil surface and the rate of rainfall. The latter is called the *rainfall intensity* (box 6.5). If the surface soil aggregates are stable, the soil will have a high initial infiltration rate, especially when it is dry. This is usually the case under a grass cover crop in the vineyard. But where the inter-rows are clean-cultivated, structural changes can occur in the surface following raindrop impact. Aggregates may slake and disperse so that small particles

Box 6.5   *Measurements of Rainfall and Rainfall Intensity*

Consider a volume of rain V (m³) falling on a soil area A (m²) in time t (hr) (fig. B6.5.1). If none of this rain infiltrates the soil but collects on the surface, we can calculate the depth d of this "ponded" rain as

$$d \text{ (mm)} = \frac{\text{Volume } V \text{ (m}^3\text{)}}{\text{Area } A \text{ (m}^2\text{)}} \times 1000 \qquad \text{(B6.5.1)}$$

Thus, the amount of rain is usually measured as a *depth* of water in mm. Nonmetric units, such as inches (in.), are used in the United States (see appendix 15).

Rainfall is measured in a rain gauge (fig. B6.5.2). Because of wind turbulence, the amount of rain collected decreases with the height of the rain gauge rim aboveground. For this reason, a rain gauge should be installed at a standard height (the rim 30 cm above the ground in Australia). Vegetation (preferably grass) around the gauge should be kept trimmed. No obstacles, like trees or buildings, should overshadow the gauge, and its position should not be too exposed to wind.

A *manual* gauge is read daily at a standard time (usually 9 A.M.). The rain collects in a transparent cylinder that is calibrated to give the volume of rain per unit area. The rainfall intensity is measured with a recording rain gauge. This is a gauge with two small "tipping buckets" and an electrical device for recording the time at which each bucket tip occurs, connected to a data logger. A bucket may tip for every 0.1 or 0.2 mm of rain that falls, which gives a very accurate measure of the rainfall intensity (mm/hr).

Figure B6.5.1 Water fluxes into and out of a defined volume of soil.

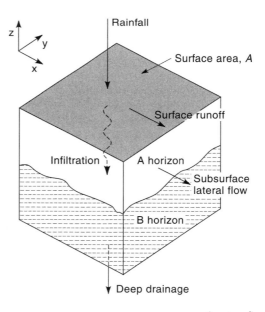

(continued)

Box 6.5 *(continued)*

Figure B6.5.2 A standard rain gauge.
Photograph by the
author.

are washed into the soil pores and cause blockages. The swelling of aggregates in a montmorillonitic clay soil as it wets up may also close off conducting pores. The result is that the soil's infiltration rate falls from its initial value to a lower rate over a period ranging from a few minutes to several hours (fig. 6.4). Even in a nonswelling soil of stable structure, the infiltration rate drops slowly over time because the gradient in ψ_m becomes smaller and smaller as the soil wets up, and gravity is the only driving force for water intake.

Figure 6.4   The time course of infiltration, surface ponding, and runoff during rainfall (White 1997). Reproduced with permission of Blackwell Science Ltd.

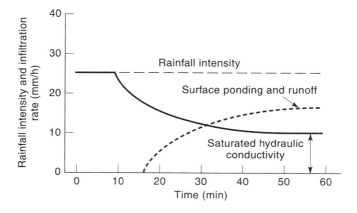

Box 6.6   *Infiltration Rate*

The *infiltration rate IR* is expressed as a volume of water/area of soil surface/time (e.g., mm/hr). When the soil's potential *IR* is greater than the rainfall intensity, all the rain should infiltrate and the actual *IR* is then set by the rainfall intensity. The cumulative infiltration *I* under this condition is given by

$$I \text{ (mm)} = IR \times t \tag{B6.6.1}$$

However, the soil's potential *IR* usually changes during a rainfall event (see fig. 6.4). If rain continues to fall at a steady rate, the soil's actual *IR* may become less than this rainfall rate, and surface runoff may occur. As with rainfall, the *runoff rate* is expressed as a volume/area of surface/time in units of mm/hr. Subsurface lateral flow, if it occurs, is calculated the same way.

Suppose that the surface runoff rate is R_s (mm/hr) from an area *A* (ha) of a vineyard block. The total volume V_s (m³) of water discharged in time *t* (hr) from the block is given by

$$V_s = \frac{R_s \times A \times t \times 10,000}{1,000} \tag{B6.6.2}$$

If the area of the block is 1 ha and the runoff period 1 hr, the volume discharged is simply $10R_s$ m³. The conversion to American units of acre-feet is given in appendix 15.

When the infiltration rate falls below the rainfall intensity, ponding occurs on the soil surface. Initially, this ponding may be confined to small depressions, which creates favorable conditions for preferential flow down macropores (section 6.3.3). But once the surface detention reaches a critical value, *surface runoff* occurs (box 6.6). In soils that have a duplex profile, downward water movement may be impeded at the top of a relatively impermeable B horizon. A zone of transient saturation can then develop (called a *perched watertable*), and water begins to flow laterally. This is referred to as *subsurface lateral flow* (box 6.5). Clearly, the speed of both surface and subsurface flow depends on the resistance offered to the flowing water and on the slope of the land, which determines the head gradient.

6.3.2   *Water Redistribution in the Soil Profile*

The velocity of water flow through a cylindrical pore varies with the square of the pore radius (i.e., r^2). Hence, we would expect water to move 100 times faster through water-filled pores of 1-mm radius than through pores of 0.1-mm radius (other factors being the same). However, factors such as the tortuosity of pore pathways and their surface roughness can vary and influence the rate of flow. Also, a soil has a range of pore sizes, so that as the soil dries, the larger, more conductive pores drain of water, and the average radius of pores still conducting water decreases. Water columns in the soil become discontinuous, so the pathways for water movement become more tortuous. Overall, the rate of water flow through the soil decreases markedly as the water content decreases.

These effects are expressed quantitatively by Darcy's Law, which relates the rate of water flow to the driving force for water movement through a proportionality coefficient called the *hydraulic conductivity K* (box 6.7).

Box 6.7   ***Darcy's Law and the Soil's Hydraulic Conductivity of a Soil***

Darcy's Law is written as

$$J_w = -K \frac{dh}{dx}$$

(B6.7.1)

where J_w is the volume of water per unit time crossing an area perpendicular to the flow and is called the water flux density, or simply the *water flux*; $\frac{dh}{dx}$ denotes the gradient or change in hydraulic head per unit distance in the direction of flow; and the coefficient K defines the *hydraulic conductivity,* which is the reciprocal of the flow resistance. (The negative sign accounts for the fact that flow occurs in the direction of decreasing head so that the gradient in h is negative). By convention, the symbols x and y denote distance in a lateral plane (usually horizontal), and z denotes distance in the vertical direction (see fig. B6.5.1).

For vertical infiltration in unsaturated soil (e.g., the beginning of an infiltration event), using head units, we may write equation B6.7.1 as

$$J_w = -K(h) \left(\frac{d(-p - z)}{dz} \right)$$

(B6.7.2)

Observing the convention for signs (section 6.1.2), this equation becomes

$$J_w = -K(h) \left(\frac{d(-p)}{dz} + 1 \right)$$

(B6.7.3)

From these equations, we note the following points:

- For unsaturated flow, where the soil water content changes with time, the value of K changes with θ and hence with the head h; that is, K is a function of h. Examples of the change in K with h in soils of different texture are shown in figure B6.7.1.

Figure B6.7.1 Change in hydraulic conductivity K of a clay soil and sandy soil with change in matric head (-p) (log scales).

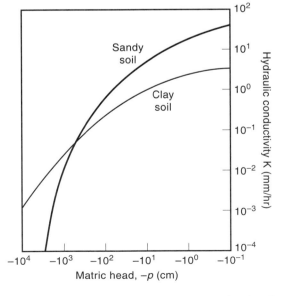

(continued)

Box 6.7 *(continued)*

- The head gradient is made up of matric ($-p$) and gravitational (z) components. When the wetted zone is deep, the matric head gradient $d(-p)/dz$ becomes small relative to the gravitational head gradient of 1, so the overall flux is downward at $J_w \cong -K(h)$.
- Suppose that after rain, the soil surface dries due to evaporation and the matric head gradient becomes negative and its absolute value >1. The product of $-K(h)$ and the head gradient then becomes positive, and water will flow upward to the surface.
- If the soil becomes saturated during rain (except perhaps for a few pockets of trapped air), the matric head gradient becomes zero and equation B6.7.3 becomes

$$J_w = -K_s \tag{B6.7.4}$$

K_s is called the *saturated hydraulic conductivity*, which is attained in the final stages of infiltration, as illustrated in figure 6.4. The value of K_s ranges from <1 to 100 mm/hr, depending on the soil's structure and texture. K_s can be measured in the field using a *ring infiltrometer*, and the variation in K for small decreases in matric head near saturation can be measured with a *disc permeameter*. More details are given in appendix 8.

6.3.3   *Water Movement and the Wetting Front*

Consider a soil that is structurally homogeneous and has no sharp changes in texture with depth. Examples are some of the deep loamy sands under vineyards in the Sunraysia and Riverland regions of the Murray Valley, Australia, and some of the deep alluvial soils of river valleys in Chile. Suppose such a soil is being wet up from a dry state by rain or irrigation. As the water penetrates more deeply, a zone of uniform water content, the *transmission zone*, develops behind a narrow wetting zone and well-defined *wetting front*. A graph of water content θ against depth z, plotted after several hours' infiltration, shows a sharp change in θ at the wetting front (fig. 6.5a). This occurs because water at the wetting front takes up a preferred position of lowest ψ in the narrowest pores, for which the hydraulic conductivity is very small. Water does not penetrate further at an appreciable rate until the large pores begin to fill.

Figure 6.5a shows that the water content in the transmission zone is approximately constant. This is described as the *steady state* condition (as in the final stage of infiltration in fig. 6.4). The condition when a soil is wetting up (or drying down) is described as non–steady state or *transient*. At steady state, and for an infiltration rate J_w, the average *pore water velocity* $\bar{v}$ in the wet soil zone is given by

$$\bar{v} = \frac{J_w}{\theta} \tag{6.7}$$

Equation 6.7 demonstrates that water entering 1 m² of soil surface is only able to travel through the wetted pore volume, defined by θ. The smaller the value of θ, the faster a given volume of water must travel through the pore space and

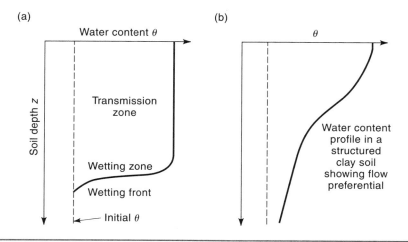

Figure 6.5 (a) The wetting profile in an initially dry soil of uniform texture and little structure. (b) The wetting profile in an initially dry, structured clay soil, with preferential flow.

hence the higher the value of $\bar{v}$. We speak of an average pore water velocity because the conducting pores are not all the same radius, and velocity is very dependent on pore radius. The distance z traveled by the wetting front in time t is given by

$$z = \frac{J_w t}{\theta} \tag{6.8}$$

This equation can be used to estimate the depth to which water has penetrated for a cumulative amount of irrigation ($J_w t$ in mm). This is also the depth to which surface-applied soluble fertilizer could move under irrigation.

In practice, soils are usually not structurally homogeneous because of the presence of old root channels and worm holes (macropores), and gaps between large aggregates. The presence of macropores and cracks predisposes to the *preferential* flow of water through the soil profile. Some of the infiltrating water flows rapidly down the macropores and fissures (referred to as "bypass flow"), whereas the bulk of the water moves more slowly into the micropores within aggregates (referred to as "matrix flow"). As a result, the wetting zone is spread over a considerable depth and the wetting front penetrates more deeply than predicted from equation 6.8 (fig. 6.5b). If preferential flow occurs, the depth of wetting can be estimated using equation 6.7, provided the fractional volume (θ_m) of the macropores is used, instead of θ for the whole soil. Some examples of average pore water velocities and depths of wetting for different soil types are given in table 6.2.

The steady state condition applies well during the winter dormant period of grapevines, especially in cool climates where evaporation rates are low. But this is rarely the condition in the root zone of actively growing vines. Intermittent rainfall, irrigation, surface evaporation, and variable water uptake (through transpiration) cause θ to fluctuate with time at any particular point in the soil. Water flow can still be described by Darcy's Law, provided it is combined with an equation for the conservation of water mass. The solution of this equation is beyond the scope of this book, but it is discussed in texts such as Jury et al. (1991).

Table 6.2   **Average Pore Water Velocities and Wetting Depths in Soils**
of Different Texture and Structure During Infiltration

Soil Type	Rainfall or Irrigation Intensity (mm/hr)	Average Effective Water Content θ During Infiltration (cm³/cm³)	Average Pore Water Velocity ν (mm/hr)	Depth z of Wetting Front in 5 hr (cm)
Sandy soil	10	0.25[a]	40	20
Clay soil with a massive structure[b]	10	0.45[a]	22	11
Well-structured clay soil with macroporosity	10	0.1[c]	100	50

[a]Representative values of θ at field capacity
[b]The infiltration rate of this soil may fall below the rainfall intensity so that ponding could occur. The depth of wetting would then be less.
[c]The macropores create a preferred pathway for water movement, and much of the soil matrix is bypassed.

6.4   *The Hydrologic Cycle in a Vineyard*

6.4.1   *Precipitation*

Rainfall, infiltration, runoff, and drainage are all part of the cycling of water among the atmosphere, soil, plants, and water bodies—the *hydrologic cycle*. The full range of processes in the hydrologic cycle for a vineyard in leaf is illustrated in figure 6.6.

Because rain may turn to hail or snow, the input from the atmosphere is often referred to as *precipitation P*. Much of the rain that falls is intercepted by the vine canopy, which may cover up to 80% of the soil surface, depending on the trellis system and vine vigor. Rain that falls in the inter-row area may be intercepted by a cover crop or fall directly on the soil. Particularly near the sea and in river valleys, the canopy can also be wetted by mist and fog—this is called *interception*. The *canopy storage capacity* amounts to 1–2 mm of water, depending on the foliage density and architecture of the canopy. Any excess water drips from the canopy or runs down the trunks to the soil. Water retained by the canopy is lost by evaporation and is referred to as *interception loss*; this can be a significant fraction of the total rainfall (up to one-quarter) if frequent light showers fall during the summer. Canopy drip, stem flow, and direct rainfall on the soil together comprise the *net precipitation* or *net rainfall*.

6.4.2   *Evaporation*

Of the rain falling on the soil, some infiltrates and some may run off, as discussed in section 6.3.1. Infiltrated water replenishes the soil water. The main loss of water from soil is by *evaporation*, which refers to loss of water in the vapor phase. Evaporation can take place from the soil surface or from leaves.

Evaporation from the soil surface. As water evaporates, a gradient in ψ_m is established between the surface and the wetter soil below, which draws water up-

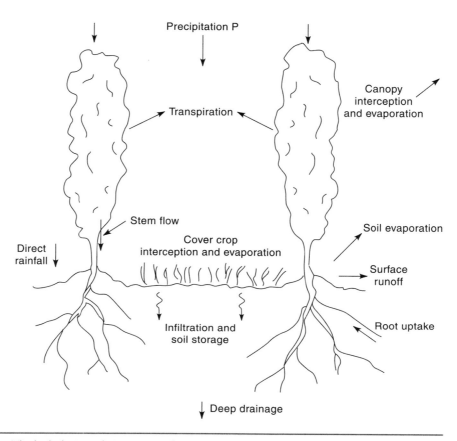

Figure 6.6   The hydrologic cycle in a vineyard.

ward. However, as the top 1–2 cm of soil dries, the hydraulic conductivity falls to a low value so that further movement of liquid water to the surface is very slow. The soil surface becomes *air-dry*. The water content of an air-dry sandy soil might be only 2–4%. This natural process of surface drying conserves water deeper in the soil profile and is called a *self-mulching effect*. The effect is enhanced by shallow cultivation of the soil to break the continuity of pores that can conduct water to the surface.

Evaporation from leaves or transpiration. Vines in full leaf provide a much larger surface than the soil for evaporation of water. Whether in the plant or soil, the energy required to convert liquid water to water vapor is provided from the atmosphere. Evaporation takes place from the wet cell walls inside a leaf, with water vapor escaping through the stomata. As a result of this evaporation, the water potential in the leaves falls as low as -1.0 to -1.5 MPa during the day. A gradient in ψ is created between the leaves and the roots, which causes water to be sucked into the roots. Water flows into the water-depleted zones around absorbing roots to maintain the transpiration stream in the vine. As long as energy is supplied to the canopy and a significant fraction of the roots is in moist soil, the vine transpires at approximately the *potential rate of evaporation E_p*. However, when the water supply becomes limiting—through depletion of soil reserves or simply due to the slow rate of flow through dry soil to the roots—the transpiration rate falls. When the vine leaves no longer maintain turgor, the vine is under severe stress and temporary wilting occurs. Turgor is restored if the vine is watered or if the evaporative power of the atmosphere falls to a low level (e.g., at night). If the dry soil condition persists, the vine may wilt permanently, which is fatal for the vine. The combination of transpiration from the plant and evaporation from bare soil is usually referred to as *evapotranspiration ET.*

In reality, there is a complex interaction among soil properties (e.g., the θ versus ψ_m relationship), plant properties (e.g., canopy structure, root density and distribution, and the osmotic adjustment of cells through salt uptake), and atmospheric factors, which affect the *ET* rate. The strong link between atmospheric factors and the transpiration rate, which determines the rate at which the plant must withdraw water from the soil to remain turgid, focuses attention on physically based measurements of evaporation, as discussed in box 6.8.

Box 6.8   *Measurement of Evaporation*

One of the simplest ways to measure evaporation is with a shallow tank of water exposed to wind and sun—an *evaporation pan.* An example is the widely used Class A pan. The daily fall in water level in the pan (allowing for any rainfall) is measured to give the pan evaporation rate E_o (mm/day). The pan should be raised above the soil surface to allow air to circulate and to minimize overheating of the water.

Another approach is to calculate the *potential rate E_p* from a vegetated land surface, using meteorological data. The basic information required is the net input of radiant energy to the system, and the capacity of the air to take up and remove water vapor. The energy balance at the land surface is discussed in section 3.5.1 and illustrated in figure 3.8. Of the absorbed daytime *net radiation R_N*, part is dissipated by evaporation, part is transferred as sensible heat to the air (J_A), and part is transferred as heat to the soil (J_H). The transport capacity of the air is estimated from the vapor pressure gradient over the surface and the average wind speed, with an empirical factor to account for surface roughness when evaporation from vegetation, rather than a free water surface, is considered.

The input variables are measured using a "weather station," as shown in figure B6.8.1. Data are collected on an hourly basis, and the evaporation rate over 24 hours is calculated in mm/day. The *reference E_p* is the rate of evaporation from a short (12 cm height) green grass crop, completely covering the soil surface, with an unlimited supply of soil water. The Penman-Monteith equation (Smith 1992) is widely used to calculate the *reference E_p*:

$$E_p = \frac{0.408\Delta(R_N - J_H) + \gamma 900 U(e_a - e_d)/(T + 273)}{\Delta + \gamma(1 + 0.34U)} \tag{B6.8.1}$$

A simpler alternative, which does not include a term for vapor pressure deficit and wind speed, is the Priestley and Taylor (1972) equation:

$$E_p = \left[k \frac{\Delta}{\Delta + \gamma} (R_N - J_H)/\lambda \right] \tag{B6.8.2}$$

(continued)

Box 6.8 *(continued)*

Figure B6.8.1 An example of a weather station in a vineyard. Photograph by the author.

In these equations, Δ is the slope of the graph of saturated vapor pressure versus temperature, γ is the psychrometric constant, U is the wind speed at a height of 2 m, $(e_a - e_d)$ is the water vapor pressure deficit, T is the average temperature (°C), and λ (lambda) is the latent heat of water vaporization. Although the dimensionless constant k should be chosen for a particular region, the value of 1.26 has been found to be widely applicable.

Depending on the surface albedo, evaporation from a wet soil surface is comparable to E_p, which is approximately equal to $0.8E_o$.

6.4.3   *Soil Water Storage*

Because the grapevine is a perennial plant that may live for more than 100 years, it can develop deep roots, especially in deep permeable soils (section 3.1). But the majority of vine roots (>70%) occur in the top meter of soil, which defines the *effective root depth*, especially for irrigated vines. The amount of water held in this root zone at field capacity *FC* defines the *soil water storage S*. As explained in box 3.5, this is most conveniently expressed as a depth of water (in mm) per meter depth of soil ($= \theta \times 1000$). For example, values of θ at *FC* for a sandy soil of 0.25 and for a clay soil of 0.45 m^3/m^3 (table 6.2) correspond to *S* values of 250- and 450-mm/m depth, respectively. Methods for measuring soil water content are discussed in box 6.9.

Box 6.9   *Measuring Soil Water Content*

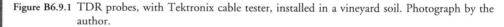

The best instruments for vineyards are those that allow repeated, nondestructive measurement of θ at several depths. One such instrument is the *neutron probe*, which depends on the emission and reflection of neutrons from a radioactive source. However, the probe does not give accurate readings within 20 cm of the soil surface, and it has the added disadvantage that the operator must have a license to use radioactive equipment.

Other techniques rely on measuring the change in bulk soil permittivity with a change in water content (water has a very high permittivity compared to soil solids and air). An electromagnetic pulse is transmitted down steel waveguides (probes) installed in the soil, and the time for the reflected pulse to return to a receiver is measured. This is the basis of *Time Domain Reflectometry* (TDR). Figure B6.9.1 shows probes installed in undisturbed soil under a vine row. The pulse generator and receiver in this case were incorporated in a "Tektronix" cable tester. Other TDRs have the generator and receiver built into the probe head. The TDR measures θ in a narrow cylinder of soil surrounding the probes.

A similar technique that relies on detecting changes in bulk soil permittivity with water content is *Frequency Domain Reflectometry* (FDR). This is the basis of the *capacitance probe*, which can be permanently installed in soil or lowered down a narrow-bore PVC tube to different depths. The orientation of the probe in the tube must be kept constant, and there should be no air gaps between the tube wall and the surrounding soil, otherwise false readings are obtained. Both TDR and FDR allow measurements to be made close to the soil surface.

More specific details of these instruments are given in Pudney et al. (2001).

Figure B6.9.1 TDR probes, with Tektronix cable tester, installed in a vineyard soil. Photograph by the author.

6.4.4   *Runoff and Deep Drainage*

The generation of surface runoff R_s and subsurface lateral flow R_{ss} is described in section 6.3.1. R_s and R_{ss} are potential losses to the vines, but they contribute to streamflow, and hence to any dams that are built to store water for irrigation. Surface runoff during rain causes a rapid response in streams that drain the vineyard area. This can be augmented by flow from *contributing areas*, which quickly reach saturation during periods of significant rainfall. Such areas can often be identified as seepage zones, fed by subsurface flow from higher areas, that occur at the base of slopes and near streams.

Water that drains below the root zone is called *deep drainage D*. Such drainage is a loss to the vines, but it may contribute to groundwater. Lateral flow of groundwater is much slower than runoff and subsurface lateral flow. It provides the bulk of the steady *base flow* of streams.

6.4.5   *The Water Balance Equation*

An area of vines that makes up a discrete hydrologic unit in the landscape is called a *catchment* (or a *watershed* in North America). The combined effect of all the hydrologic processes in a catchment can be expressed through the *water balance equation*. This equation assumes a mass balance of water, so that

$$P = E_a + R_s + R_{ss} + D + \Delta S \qquad (6.9)$$

All the terms in this equation are in mm. E_a is the actual evapotranspiration from the soil–plant system, and ΔS (delta S) is the change in soil water storage. Typically, in winter rainfall areas when the vines are dormant, the soil should be at or close to *FC*. If the vineyard soil is well managed (adequate infiltration and good structure), R_s and R_{ss} should be small to negligible. Hence, when ΔS is small, equation 6.9 simplifies to

$$P = E_a + D \qquad (6.10)$$

Because E_a is small during winter, drainage below the root zone during winter can be significant. But from spring to early summer, the soil water content starts to fall as the vine canopies develop and evapotranspiration increases. During summer, D is insignificant and equation 6.9 simplifies to

$$P = E_a + \Delta S \qquad (6.11)$$

When E_a exceeds P for a period, ΔS in equation 6.11 becomes negative and a substantial *soil water deficit* (*SWD*) develops. This *SWD* represents a withdrawal from the soil's *plant available water* (*PAW*) which depends on the *AWC* in mm/m depth of soil (section 6.2.2) and the rooting depth of the vines in meters (section 7.1.3).

6.5   *Managing the Soil Water Supply*

Good vineyard management involves, as far as possible, managing the soil water supply. Shoot growth indicates the vine's water and nutrient status. Well-watered vines have actively growing tips and normal internode expansion. If water is too readily available, especially during the period of rapid shoot growth from pre-flowering to veraison (see fig. 5.6), excess vigor may result.

However, if water stress occurs, shoot growth slows, internodes shorten, and the tip becomes a dull grey-green. The leaves become flaccid. With severe stress, growth stops and the tips and tendrils may die. Late in the season, water-stressed lower leaves become yellow and may develop necrotic areas at the edges; leaf drop begins at the basal leaves and progresses toward the tip.

Severe stress at flowering reduces fruit set. From fruit set to veraison, berry size is decreased by moderate to severe stress (an effect on cell division). During ripening, mild stresses can enhance the accumulation of soluble solids by suppressing vegetative growth. Under more severe stress, soluble solids increase through a reduction in berry size or shriveling (an effect on cell expansion). In this case, sugar accumulation and flavor development may be delayed because of reduced photosynthesis and premature leaf drop.

In dryland vineyards, several factors affect the soil water supply to the vines:

- Slope of the land and whether it is contoured or terraced (section 7.5.4)
- Infiltration rate of the surface, which depends on soil type, the presence of a cover crop, and the extent of compaction from wheeled traffic (section 7.1.1)
- Type of cover crop (annual or perennial, and its depth of rooting) and weeds in the vine rows, all of which compete with the vines for water (section 7.3.1)
- Mulching with straw or cover crop residues in the vine row to reduce soil evaporation (section 7.3.2)
- Installation of subsoil drainage to remove surplus water and control the height of the water table (section 7.2.4)

All these factors can also operate in an irrigated vineyard; but the availability of irrigation water, and manipulation of its amount and timing, enable a higher degree of control of the soil water supply, especially in environments with low rainfall. There are several approaches to managing irrigation to meet specific production objectives.

6.5.1   *Using Soil Water Potentials to Manage Irrigation*

Based on the soil water retention curve (fig. 6.3b), the *available water capacity* AWC can be subdivided into readily available water (RAW), deficit available water (DAW), and the remainder ($AWC - RAW - DAW$). With a FC corresponding to $\psi_m = -10$ kPa, the lower limit of RAW is set between -40 and -60 kPa. Grapevines can obtain this water without any stress. The lower limit of DAW is set in the range -60 to -400 kPa, which defines the *stress limit* for the vines; the vines can extract water held at lower potentials, but the rate of supply is too slow for adequate physiological functioning.

The stress limit varies with the age of vines, soil type, root distribution, and weather conditions. For example, for young vines (≤ 3 years) growing in a sandy loam soil under hot summer conditions, the stress limit should be set at -60 kPa; for older vines under the same conditions, the stress limit can be set at -100 kPa. On the other hand, for old, deep-rooted vines growing in a clay loam soil, or a soil at least 1 m deep over permeable limestone under cooler conditions, the stress

Figure 6.7   A pressure chamber for measuring leaf water potential. Photograph by the author.

limit is more likely to be -400 kPa. These stress limits have been identified by experiments in which ψ_m values at different depths in the root zone have been compared with the leaf water potential (ψ_l) in the vines. The ψ_l value is measured on individual leaves plucked from the vine at a particular time of day and placed in a pressure chamber (fig. 6.7). The ψ_l of a nonstressed vine, measured at midday, should not be <-1 MPa (-1000 kPa), whereas that of a stressed vine can fall as low as -1.5 MPa (Williams 1996).

Because measurements are made at a point (see box 6.3), ψ will vary with depth and distance from each vine, depending on the root distribution and the balance between rainfall and ET. Values of ψ_m are also influenced by soil structure and texture, which vary spatially. Tensiometers are commonly used under young vines that are kept well watered ($\psi_m > -60$ kPa); otherwise, gypsum blocks are used. These instruments should be installed at a minimum of two depths (the midroot zone at 25–50-cm depth, and near the bottom at ca. 1-m depth). Because ψ as measured will change with depth as a result of the change in ψ_g, the measured values must be corrected to obtain the ψ_m value (see equation 6.4). Tensiometers or gypsum blocks should be placed in the row (so as not to obstruct between-row operations) and between vines, at a suitable distance from any drip irrigation emitter or minisprinkler (10–15 cm in a sandy soil and 20–25 cm in a clay soil).

Growers do not want their vines to suffer water stress before flowering and at the time of flowering (ψ_m in the midroot zone should not fall below -40 kPa in sandy soils and -60 kPa in clay soils). Between flowering and veraison, the grower may want to impose some stress to moderate shoot growth and slow down cell division in the berries (to control berry size). Hence, ψ_m (midroot zone) should not fall below about -100 kPa in sandy soils, -200 kPa in loamy soils, and -400

Table 6.3   ***Typical Values for RAW and DAW in Soils of Different Texture***

	Loamy Sand to Sandy Loam	Sandy Clay Loam to Clay Loam (mm/m Depth of Soil)	Silty Clay to Clay
RAW	50	70	60
DAW	30	50	70
RAW + *DAW*	80	120	130

kPa in clay soils. When these limits are reached, irrigation should be applied. The difficulty is to know how much water to apply to return the bulk of the root zone to -40 to -60 kPa, an issue discussed in the following section.

6.5.2   *Using Soil Water Deficits to Manage Irrigation*

The ψ_m range between no stress (-40 kPa) and the lowest stress limit (-400 kPa) defines the desirable deficit range. Because of the relationship between θ and ψ_m, these stress limits can be expressed in terms of *SWD*. The control of *SWD* in a desired range by irrigation is called *regulated deficit irrigation* (*RDI*).

To implement *RDI*, the value of θ should be measured at regular intervals down to at least 1 m, so that the value of *S*, in mm of water per m depth of soil, can be calculated (section 6.4.3). At least one site per vineyard block should be chosen, in a way that samples any variation in profile soil water content caused by soil type and topography. By regular measurement of *S*, the development of a *SWD* can be monitored. The *SWD* is then related to the soil's *RAW* and *DAW*, and a decision can be made on when to irrigate and how much water to apply. Examples of *RAW* and *DAW* values for soils of different texture are given in table 6.3. An example of *RDI*, based on this approach, is given in box 6.10.

6.5.3   *Using Crop Factors or Crop Coefficients to Manage Irrigation*

Rather than directly measuring the *SWD*, we can calculate the expected *SWD* from weather data. In most vineyards, the soil will be at *FC* at bud burst. The subsequent development of an *SWD* depends on the balance between rainfall and evapotranspiration, and the rooting volume of the vines.

Rainfall is easily measured (box 6.5), and *ET* can be estimated from an evaporation pan (E_o) or calculated from weather data (E_p). However, the *actual evaporation rate* (E_a) from a vineyard depends on the vine spacing and canopy development, and the degree of soil water stress imposed on the vines. If E_o values are available, an empirical *crop factor* C_f is used to calculate E_a (mm/day), according to the equation

$$E_a = E_o \times C_f \tag{6.12}$$

An example of crop factors for a vineyard in southern Australia is given in table 6.4. These factors are for vines that are well watered or under *RDI*. When the vines are experiencing some stress (under *RDI*), the transpiration rate is lowered and the crop factors are therefore less than for fully irrigated vines. If E_p is

Box 6.10   *Regulated Deficit Irrigation and Irrigation Scheduling*

First, calculate the stored soil water S at *FC*. An example of how to do this is given in appendix 9. Suppose that the soil is a uniform sandy loam for which $S = 300$ mm/m depth of soil. From table 6.3, the *RAW* and *DAW* per m depth are 50 and 30 mm, respectively. Hence

Critical *SWD* to trigger irrigation $= -(RAW + DAW) = -80$ mm   (B6.10.1)

Since the *FC* value is 300 mm, the critical *SWD* is reached when the stored soil water is at the lower stress limit, S_{sl}:

$$S_{sl} = 300 - 80 = 220 \text{ mm} \qquad (B6.10.2)$$

The regulated deficit range for this soil is 30 mm/m depth, and the value of S_{ns} at the no-stress limit is

$$S_{ns} = 300 - 50 = 250 \text{ mm} \qquad (B6.10.3)$$

Thus, the *SWD* should fluctuate between -80 and -50 mm/m depth of soil. When the soil has reached the lower stress limit, 30 mm of irrigation should be applied to bring it back to the no-stress limit. Therefore, 30 mm is the depth of water to be applied per ha and is equivalent to 30 L/m^2 of soil area. This calculation assumes that the vine roots explore all the soil to a depth of 1 m.

Vine root distribution depends on a number of factors, the most important of which follow:

1. Vine age (the older the vine, the deeper its roots), variety or clone, and rootstock (but see factor 2)
2. Soil properties, whether there is a physical or chemical impediment
3. Method of irrigation (drip, minisprinklers, overhead sprinklers, furrow, or flood), and its scheduling (frequent and shallow versus infrequent and deep)

Factors 1 and 2 are discussed in more detail in chapter 7. Factor 3 is discussed in section 6.6. Appendix 10 gives examples of *RDI* volumes calculated for different root distributions, according to soil type, number of vines per ha, and method of irrigation.

used instead of E_o, a different factor, called a *crop coefficient* C_c, is used to calculate E_a as follows:

$$E_a = E_p \times C_c \qquad (6.13)$$

Because $E_p < E_o$, crop coefficients are larger than crop factors for the same conditions, as shown in table 6.4. The measurement of crop factors and coefficients is discussed in box 6.11.

The development of a *SWD* is calculated using equation 6.11 for a convenient time period, usually 1 week. Assuming the soil is at *FC* at bud burst (maximum stored soil water S), for week 1

$$\Delta S = P - E_a = SWD_1 \qquad (6.14)$$

and for week 2

$$SWD_2 = SWD_1 + (P - E_a) \qquad (6.15)$$

and so on.

Table 6.4   *Crop Factors and Crop Coefficients for Vines at Different Growth Stages on a Sandy Loam Soil in a Temperate Climate*

Stage of Growth	Crop factor C_f[a]		Crop coefficient C_c[b]	
	No Stress	RDI[c]	No Stress	RDI[c]
Bud burst	0.1	0.1	0.2	0.15
Flowering	0.25	0.25	0.4	0.3
Veraison	0.5	0.25	0.7	0.5
Harvest	0.5	0.25	0.75	0.55
Post-harvest	0.25	0.15	0.75	0.55

[a]The crop factors are for healthy vines, on a single trellis up to 2 m high, with a maximum canopy coverage of 70–80% of the ground, and a mown, inter-row cover crop (Goodwin 1995).
[b]The crop coefficients are for similarly managed vines, except that the inter-rows are bare. A decrease in either C_f or C_c postharvest depends on whether the vine canopy continues to actively photosynthesize.
[c]Assuming that the SWD is to be maintained between the no-stress and lower stress limits (SWD between -50 and -80 mm/m depth in this soil)

Although a weekly time period is used, rainfall and evaporation should be measured daily. A SWD develops only when $E_a > P$, and the solution to these equations is negative. SWD values calculated in this way are used to determine whether the soil water content lies in the desired range. In California's Central Valley, where the weather during the growing season is very reliable, irrigation can be scheduled in advance by substituting historical average weather data in these equations. This approach is called the "volume balance approach" (Williams

---

Box 6.11   *Measurement of Crop Coefficients and Crop Factors*

E_a for vines can be measured using a *weighing lysimeter*, similar to that installed at the Kearney Agriculture Center at Fresno, California. This lysimeter, which is 2 m wide × 4 m long × 2 m deep, is planted with two vines, but is located in a vineyard so that the ambient conditions for these vines are no different from other vines. The lysimeter, which sits on a weighing mechanism so that E_a can be measured from its change in weight, is automatically drip-irrigated when $ET = 2$ mm. Thus, the actual evaporation rate from well-watered vines can be measured as the grapevine develops, and the ratio of E_a to E_p established to obtain the crop coefficient C_c. In California, E_p values calculated from a Penman-type equation are available from a network of 90 weather stations that comprise the California Irrigation Management System (Pitts et al. 1995).

Weighing lysimeters of this size are not commonly available. However, Williams (1999) has found that vine water use at Fresno is linearly related to the percentage of soil surface shaded by the vine canopy, once this is >15%. This property, measured using a digital camera, can be used as a surrogate variable for E_a. An advantage of this approach is that differences in shaded area due to different trellis systems can be measured, and values of $E_a/E_p = C_c$ for particular vine-trellis systems calculated. Where pan evaporation E_o is measured, values of C_f are calculated the same way. RDI can also be applied to the vines in the lysimeter to enable C_c and C_f values for vines under some stress to be estimated.

1999). An example of these calculations, using crop coefficients from table 6.4 and target *RDI* values, is given in appendix 11.

6.5.4   *Partial Root Zone Drying*

Partial root zone drying (*PRD*) is a recently developed technique that shows promise for regulating vine growth and saving irrigation water, without sacrificing grape yield and quality (Dry and Loveys 1998). The physiological basis of *PRD* is that a hormonal signal is sent from vine roots to the shoot when the roots experience a shortage of water. The short-term signal appears to be an increase in abscisic acid concentration, which induces partial stomatal closure in the leaves and a reduction in transpiration. A longer-term response may be a decreased production of cytokinins, which slows shoot growth. However, the plants adjust to the drier soil conditions and the regulating effect disappears with time unless the dry soil zone is alternated. This is achieved by having a dripper on each side of a vine, and alternating the side of the vine that is irrigated every 10–15 days (fig. 6.8).

Experiments in commercial vineyards in southeastern Australia have shown that with reduced lateral shoot growth, leaf area was decreased and bunch exposure improved. In some cases, there were small increases in sugar content and titratable acidity of the fruit. Yields may be slightly decreased, but the quantity of water used during a season is as much as 50% less than that used in normal irrigation practices. *PRD* requires two dripper lines per row and sensors installed to monitor the changes in θ on either side of the vine. The savings in water are not necessarily greater than with properly managed *RDI*, as can be seen in appendix 10. Further assessment of the advantages and disadvantages of the two approaches for grape production, berry quality, and water saving is in progress for a range of soils and climates.

Figure 6.8   A young vineyard with drip irrigation lines for PRD. Photograph by the author.

6.6   *Irrigation Systems*

Irrigation systems may be broadly divided into *microirrigation* systems, made up of drip (or trickle) systems, microjets, and minisprinklers, and *macroirrigation* systems, comprising overhead sprinklers, traveling irrigators, and flood/furrow systems.

The choice of which system to use depends on several factors, including the availability and price of water, the topography and soil type in the vineyard, capital costs of installation, and the skill of the vineyard manager. However, the efficiency of water use (from pump to soil) is normally highest for drip systems (90%), intermediate for sprinklers (60–80%), and lowest for surface furrow and flood systems (<50%).

6.6.1   *Microirrigation Systems*

Details of the construction and performance of these systems are given in specialized irrigation books, such as by Dasberg and Or (1999) or Mitchell and Goodwin (1996).

Drip systems consist of emitters spaced at regular intervals along flexible PVC pipes that run along the vine rows. Individual emitters deliver water at rates between 1 and 8 L/hr. Drip lines may be placed above ground or below, the latter being called *subsurface drip irrigation* (*SDI*). For *SDI*, the lateral pipes are laid 1–2 m apart at depths of 0.4–0.5 m, depending on the soil depth and hydraulic conductivity. Many of the advantages and disadvantages of drip irrigation and *SDI* are similar (table 6.5). However, some aspects specific to *SDI* are the following:

* Root invasion and consequent blockage can occur. To prevent this, the irrigation water filters or the in-line drippers are impregnated with the herbicide trifluralin ("Treflan"). Root invasion is also minimized by keeping the soil around the drippers moist most of the time.
* If the system develops a fault, a vacuum may develop which sucks water and soil into the emitters. Vacuum relief valves should be installed at the head of the system and at high points along the laterals. The system should be flushed frequently to prevent blockages and inhibit root invasion.
* There is no wet soil surface under the vines, which reduces the risk of fungal infections, such as downy mildew, powdery mildew, and botrytis.
* Soil evaporation is reduced, hence there is no need for mulch, which can be a fire and frost hazard in some orchards.

Microjets deliver water at a rate of 32–40 L/hr compared to a rate of 50–100 L/hr with *minisprinklers*. A microjet can be directed to give a uniform band of wet soil in the vine row and can be inverted for young vines to reduce the wetted area. Minisprinklers wet a larger soil area than microjets or drippers. Wetting with a minisprinkler is more uniform than with a microjet, provided the operating pressure is adequate. If a minisprinkler is operated at pressures <150 kPa, the wetting pattern is a nonuniform "doughnut" shape. Both microjets and minisprinklers require filtered water, or water with a low concentration of suspended solids, to avoid blockages.

Neither microjets nor minisprinklers are as effective as drippers in saving water (see appendix 10).

Table 6.5   ***Advantages and Disadvantages of Drip Irrigation in Vineyards***

Advantages	Disadvantages
Water applied to individual vines can be closely controlled; particularly suited to *RDI* and *PRD*. With frequent application, soil water content (and ψ_m) in the root zone is maintained in a narrow range.	Blocking of emitters (0.5–1.2 mm orifices) due to suspended particles in the water, chemical precipitates, or microbial growths. Water must be filtered, and wastewater may need to be chlorinated. Root invasion can be a problem for *SDI*.
Very low losses by evaporation and runoff (especially for *SDI*), because area of wet soil is small and rates of application are low. No foliar salt uptake as with overhead sprinklers.	Salt accumulates at the edge of the wetted zone, especially if there is insufficient winter rain to leach these salts out of the root zone (more of a problem with low-quality irrigation water)
Very low drainage losses (and hence loss of nutrients or potential contaminants) if *RDI* is practiced.	Rodents, other mammals, and birds can damage the soft, flexible pipes when they are seeking water (not a problem for *SDI*).
Targeted fertilizer application by *fertigation*; see chapter 5.	Less effective microclimate control than with overhead sprinklers, which can be used to avoid frost damage to vines at critical times, such as bud burst, flowering, and fruit set.
Allows irrigation water of higher salinity to be used, compared with overhead sprinkler or flood systems (where irrigation is less frequent and the soil dries between applications).	More expensive than overhead sprinklers.
Wastewater can be used because there is no chance of pathogens contaminating foliage or fruit; see chapter 7.	
Restricted weed growth, especially with *SDI*, and no constraints on access because of wet soil.	
Effective on marginal soils and in difficult topography (stony soils, steep slopes).	
Low operating pressures (100–200 kPa) mean lower energy costs for pumping compared to overhead sprinklers.	

6.6.2   *Macroirrigation Systems*

A network of *overhead sprinklers* wets all the soil surface, so excess water may be applied in parts of the vineyard, which can lead to drainage losses. Losses by evaporation as the water is being applied can be high, and salts in the water can be absorbed into the leaves via stomata, which is undesirable. Access to the vines is restricted during, and for a time after irrigation. However, overhead sprinklers can be used to raise the humidity of the air when a frost is likely: the high specific heat capacity of water (section 3.5.2), and its high latent heat, help to prevent the air temperature from falling below 0°C. Overhead sprinklers need not be fixed,

Figure 6.9	A furrow irrigation system and water metering device for a vineyard in Sunraysia, Australia. Photograph by the author.

but can be part of a "traveling irrigator" that moves along the vine rows, as in the Coonawarra and Padthaway regions of South Australia.

With *furrow irrigation*, water is diverted from a main ditch generally into a channel running on either side of each row (fig. 6.9), or sometimes into a single central channel. For *flood irrigation*, the vineyard is divided into irrigation bays, and water flows into the bays to cover the entire row and inter-row area (except where soil is "mounded" along the vine row). Sufficient water must be applied for soil at the far end of the bay to wet to a desired depth, which often means that vines at the head of the bay receive too much water. This effect is exacerbated in long bays on permeable soils. The problem of uneven distribution of water is compounded if there are slight rises and depressions along the bay, because water will pond in the depressions, where infiltration will be greater. This can be overcome by leveling the bays with laser-guided graders.

Furrow irrigation is popular in older vineyards of the Lodi District in California. One advantage of furrow irrigation in this district is that water from surface storages can be used, rather than the scarcer groundwater, which, because of its lower sediment content, is preferred for drip irrigation. Flood irrigation is common in older vineyards in Chile, and also occurs in the Riverland, Sunraysia, and Griffith regions of Australia. Furrow and flood irrigation waste water compared to drip irrigation (see appendix 10). An example of poor management of flood irrigation is shown in figure 6.10.

Figure 6.10   Poor control of water applied under flood irrigation in the Riverland region of South Australia. Photograph by the author.

6.7   *Summary Points*

This chapter explored the complex relationships among soil, water, and grapevines. The major points are as follows:

■ Water is held in the soil's pore space, and the volume of water per unit volume of soil defines the *volumetric water content* θ. Various forces, adsorption, capillary, and osmotic, as well as gravity, act on soil water to reduce its free energy. The extent of the reduction in energy is measured by the *soil water potential* ψ. The effect of individual forces is expressed through the component potentials, *matric* ψ_m (negative) and *pressure* ψ_p (positive), *gravity* ψ_g, and *osmotic* ψ_s.

■ The water potential can be expressed as energy/mole, or more commonly as energy/volume (pressure p) or energy/weight of water (head h). The units of head are meters, which are most convenient when describing the movement of water through soil using Darcy's Law.

■ Darcy's Law defines the relationship between the water flux through a cross section of soil (J_w) and the head gradient (dh/dx), as

$$J_w = -K(h)\,\frac{dh}{dx}$$

■ K is the *hydraulic conductivity* (mm/hr), which depends on h. When the soil is saturated, $K = K_s$, the saturated hydraulic conductivity. K_s is the maximum value of K, and K falls sharply as the matric head ($-p$) decreases, more so in sands than in clay soils. K_s determines the steady state infiltration rate IR of a soil. When the rainfall or irrigation rate exceeds the surface IR, surface runoff may occur.

▪ The relationship between θ and ψ_m is called the *soil water retention curve*. This relationship determines how much water is released from the soil pores under a negative pressure (or suction) or how much drains under the influence of gravity. The θ value to which a soil drains, 48 hours after being thoroughly wetted, defines its *field capacity* (*FC*), when ψ_m is approximately -10 kPa. The θ value at which plants lose turgor and wilt is called the *permanent wilting point* (*PWP*), when ψ_m is approximately -1500 kPa. The amount of water held between *FC* and *PWP* is the *available water capacity* (*AWC*), which is the maximum amount of plant-available water per unit depth of soil. The pore-size distribution of a non-swelling soil can be derived from the soil water retention curve.

▪ The processes of rain falling, infiltration, runoff (by surface and subsurface lateral flow), evaporation, and drainage are part of the *hydrologic cycle* in a vineyard. Infiltrated water is redistributed according to gradients in ψ in the soil profile. The water balance of a vineyard over time may be written as

$$P = E_a + R_s + R_{ss} + D + \Delta S$$

where P is precipitation, E_a is the actual evapotranspiration from soil and vines, R_s and R_{ss} are surface and subsurface runoff, respectively, D is deep drainage below the root zone, and ΔS is the change in soil water storage. All these terms are measured in mm, which is a volume of water per unit area. When runoff and drainage are negligible and $E_a > P$ (as in summer), ΔS is negative and a *soil water deficit* (*SWD*) develops.

▪ When vines are well supplied with water, they transpire water (evaporation through the leaves) at the *potential rate E_p*, which is determined by the net solar radiation absorbed and the weather conditions. As soil water becomes limiting, the vine begins to suffer some water stress and the transpiration rate falls. Soil water held at ψ values above -40 to -60 kPa, depending on soil texture, is called *readily available water* (*RAW*), whereas water held from -40 to -60 kPa down to -100 to -400 kPa is called *deficit available water* (*DAW*).

▪ *RAW* and *DAW* should be calculated for the effective rooting depth of the vines, which in an irrigated vineyard may be 60–70 cm. Managing the soil water within these ranges is the basis of *regulated deficit irrigation* (*RDI*). *RDI* is important for controlling "excess vigor" in irrigated vines, especially in warm to hot regions. *Partial root zone drying* (*PRD*) is another technique for saving water, without sacrificing yield or grape quality. Dryland vines usually have more extensive root systems than irrigated vines.

▪ *RDI* can be applied by monitoring soil ψ values or by calculating the *SWD*, either by direct measurement of θ using a range of instruments or by estimating E_a. Where E_p is known, E_a is the product of E_p and C_c, a crop coefficient, which depends on the stage of canopy development and the degree of water stress to be imposed. If only pan evaporation E_o is measured, E_a is the product of E_o and a crop factor, C_f.

▪ Irrigation by *overhead sprinklers* or *traveling irrigators* and by *flood* or *furrow* wets all or most of the soil, whereas *drippers, microjets,* and *minisprinklers* allow targeted application and conserve water. Drip irrigation is most suited to *RDI* or *PRD*, but the water should be filtered to avoid blockage of emitters. Because the wet soil zone is at a near-constant water content, water of higher salinity can be used, but adequate winter rain is necessary to leach any accumulated salts at the edge of the wetted zone. The efficiency of water use is higher for drip systems than for other systems.

7   *Soil Quality in Vineyards*

7.1   *The Physical Environment*

The soil must provide a favorable physical environment for the growth of vines—their roots and beneficial soil organisms. Some of the important properties contributing to this condition are infiltration rate, soil strength, available water capacity, drainage, and aeration.

7.1.1   *Infiltration Rate*

Ideally, the *infiltration rate IR* should be >50 mm/hr, allowing water to enter the soil without ponding on the surface, which is predisposed to runoff and erosion. The range of infiltration rates for soils of different texture and structural condition is shown in table 7.1. Typically, the soil aggregates should have a high degree of water stability so that when the soil is subjected to pressure from wheeled traffic or heavy rain, the aggregates do not collapse, nor do the clays deflocculate. Some of the problems associated with the collapse of wet aggregates and clay deflocculation, and the formation of hard surface crusts when dry, are discussed in section 3.2.3. Pans that develop at depth in the soil profile, as a result of remolding of wet aggregates under wheel or cultivation pressure, can be barriers to root growth.

7.1.2   *Soil Strength*

Soil strength is synonymous with *consistence*, which is the resistance by the soil to deformation when subjected to a compressive shear force (box 2.2). Soil strength depends on the soil matrix potential ψ_m and bulk density BD, as illustrated in figure 7.1. In situ soil strength is best measured using a *penetrometer*, as discussed in box 7.1. The soil strength at a ψ_m of -10 kPa (*FC*) should be <2 MPa for easy root penetration and should not exceed 3 MPa at -1500 kPa (*PWP*). As shown in figure 7.1, when ψ_m is between -10 and -100 kPa, the soil strength

173

Table 7.1   ***Infiltration Rates* IR *for Soils of Different Texture and Structural Condition***

IR at Steady State (mm/hr)	Grading of the IR	Typical Texture and Structural Condition
0–10	Very poor	Heavy clay soils with poor structure (often degraded by cultivation)
10–20	Poor	Clay soils with moderately good structure
20–50	Moderate	Loam, clay loam soils, some sandy soils, usually good structure
>50	Good	Loam to sandy soils with very good structure; clay or clay loam soils of the Krasnozem type

increases with BD. The BD of vineyard soils can increase, particularly in the inter-row areas because of compaction by machinery, such as tractors, spray equipment, and harvesters. Typically, compaction occurs at depths between 20 and 25 cm and is more severe in sandy soils than in clay loams and clays (except when the clays are sodic; see section 7.2.3). Figure 7.2 shows the marked difference in soil compaction, measured by penetration resistance, under a wheel track and under a vine row on a sandy soil in a vineyard.

7.1.3   *Available Water Capacity*

The *available water capacity* (AWC) is calculated from the difference between the soil's water content θ at field capacity FC and permanent wilting point PWP (section 6.2.2). The AWC, expressed in mm water per m depth, depends primarily on the soil's texture and structure. However, the *plant available water* (PAW) depends on both the AWC and on the rooting depth of the plant. A rooting depth of at least 60 cm is desirable for irrigated vines, and up to 2 m for mature non-

Figure 7.1   The relationship between soil strength and soil matric potential ψ_m.

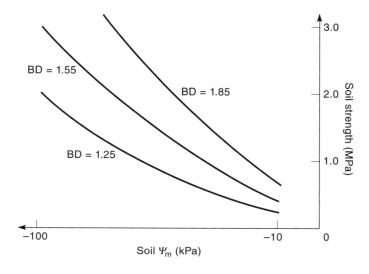

Box 7.1   *Measuring Soil Strength in the Field*

The soil's strength is measured by the strength of the resisting force it offers to penetration with a *penetrometer*. This instrument consists of a steel rod about 1 m long, with a conical tip at the base and a proving ring and strain gauge at the top (figure B7.1.1). As the conical tip is pushed into the soil, the ring is distorted and the force recorded by the gauge. The rod is graduated so that the depth of penetration can be measured. The readings are most reliable when the rate of entry of the rod is constant. Penetration resistance can be very variable spatially, so a large number of readings are required. Because soil strength depends on soil wetness (see fig. 7.1), the measurements should be made at the same water content, usually the *FC*. Penetrometer measurements are used to detect natural or induced hard pans in soil, and hence to guide the deployment of ameliorative treatments, such as deep ripping.

Figure B7.1.1 A cone penetrometer for measuring soil strength (White 1997, courtesy of D.B. Davies). Reproduced with permission of Blackwell Science Ltd.

irrigated vines. Table 7.2 shows some typical *PAW* values for soils of different texture for vines with several rooting depths.

However, the *PAW* is only a measure of the potential size of the "soil water bucket" that can hold available water. More important for quality grape production is the avoidance of undue water stress at critical stages of flowering and fruit development. In the absence of irrigation, this requires the soil to release a sub-

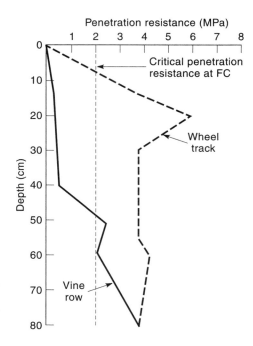

Figure 7.2   The effect of wheeled traffic on soil compaction in the inter-row of a sandy soil (after Myburgh et al. 1998).

stantial amount of water at relatively high ψ_m values. The concept of readily available water (*RAW*)—water released between -10 kPa and -40 to -60 kPa—was introduced in section 6.5.1. A *RAW* value >70 mm per m depth is considered most desirable.

7.1.4   *Drainage and Aeration*

Good drainage and aeration are usually complementary soil attributes. A soil should drain rapidly from saturation to field capacity, at which point all the macropores are air-filled. The resultant air-filled porosity or *air capacity* (see box 3.6) should be at least 10–15% to allow an adequate rate of O_2 and CO_2 diffusion.

Table 7.2   *Examples of* PAW[a] *in Soils of Different Texture, Adjusted for the Rooting Depth of Grapevines*

Rooting Depth (m)	Soil Texture		
	Loamy Sand	Sandy Loam Over Clay (Duplex Profile)	Uniform or Gradational Clay Loam
1.0	100	150	180
1.5	150	225	270
2.0	200	300	360
3.0	300	450	540

[a]Expressed as mm of water in the soil profile

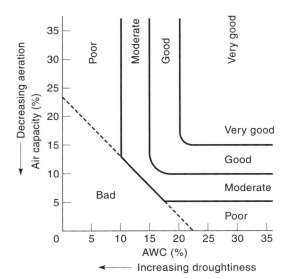

Figure 7.3 A classification of structural quality for vineyard soils (White 1997). Reproduced with permission of Blackwell Science Ltd.

Values of *AWC* and air capacity have been used to create a classification of *structural quality*, in terms of the soil's susceptibility to waterlogging and droughtiness (fig. 7.3). Originally developed for topsoils in England and Wales (Hall et al. 1977), this classification has proved valuable for vineyard soils (Cass et al. 1998) and sets practical ranges for good aeration and water availability.

7.2 *The Chemical Environment*

A satisfactory chemical environment requires an adequate supply of macro- and micronutrients, and an absence of any element toxicities and undesirable residues from pesticides. Soil nutrients and their supply to vine roots are discussed in chapters 4 and 5. Key management practices for achieving a satisfactory chemical environment are discussed in this chapter.

7.2.1 *Soil and Plant Testing*

For vineyards, the most important elements to monitor are N, P, K, and the micronutrients, usually by plant analysis (section 5.2.1). Appropriate decisions on fertilizer application can then be made. Soil testing to a depth of 1 m, noting any marked soil horizon changes, should be carried out when a vineyard is established, as discussed in section 8.2.2.3.

Testing for soil pH is important because of its influence on the availability of micronutrients (section 5.5.2) and also on the concentration of potentially toxic Al^{3+} ions. Cass (1998) recommends a pH range (in water) of 5.5–8 for vines. Vine root hairs, which facilitate the uptake of immobile nutrients such as P, are most abundant at pH 5.7 (Richards 1983). Below pH 5.5, exchangeable Al^{3+} may exceed 1 mg/kg soil, and thus adversely affect root growth. However, because grapevines grow successfully on low pH soils in many parts of the world, Winkler et al. (1974) suggest that low pH itself is not critical. For example, in the delta area of the Central Valley, California, grapes are grown on soils of pH 3.4–4.5. But a high organic matter content of 10–20% may ensure that most of the Al^{3+} ions mobilized at such a low pH are complexed by organic ligands, which renders them harmless to the roots. Liming to raise the soil pH is discussed in section 5.5.3.

Other important chemical indicators of the quality of the soil environment are *salinity* and *sodicity*.

7.2.2   *Soil Salinity*

Salinity refers to the total concentration of dissolved salts in a given volume of soil. The major ions contributing to salinity in the soil and in water used for irrigation are Ca^{2+}, Mg^{2+}, Na^+, Cl^-, $SO_4{}^{2-}$, and $HCO_3{}^-$, with low concentrations of K^+ and $NO_3{}^-$. Salinity is most conveniently measured by the *electrical conductivity* (*EC*) of the medium—bulk soil, a soil extract, or a water sample.

7.2.2.1   *Water Measurements*

EC is the *specific conductance* of the water, independent of sample size, and is measured with a conductivity meter. The total concentration of dissolved salts (*TDS*), sometimes referred to as the "salinity hazard," is related to *EC* by the equation

$$TDS \text{ (mg/L)} \cong 640 \, EC \text{ (dS/m)} \tag{7.1}$$

This empirical equation, developed from the analysis of many surface and groundwaters, holds for water with an *EC* up to 10 dS/m. Note that *EC* increases with temperature and should be corrected to a standard temperature of 25°C. In Australia, the *EC* of irrigation water is commonly given in "EC units," which are μS/cm (1 dS/m = 1000 μS/cm).

In the western United States, *EC* classes of <0.25, 0.25–0.75, 0.75–2.25, and >2.25 dS/m at 25°C have been used to define waters of *low, medium, high*, and *very high* salinity. The *EC* of most irrigation waters lies between 0.15 and 1.5 dS/m. In soil, however, salts from the irrigation water become more concentrated due to evapotranspiration. With time, the soil salinity may reach levels that are detrimental to vine growth, unless a salt balance is achieved through irrigation management (section 7.2.2.4).

7.2.2.2   *Soil Measurements*

The bulk *EC* of soil depends on both the conductivity of salts in solution and the surface charges on clays and organic matter. The surface charges are immobile and constant (in the short term), so the concentration of dissolved salts determines soil salinity. A standard laboratory method of measuring *EC* involves mixing a sieved soil sample with deionized water to make a "glistening paste," then extracting the solution by vacuum filtration. The *EC* measured on this *saturation extract* is called the EC_e. *EC* is also measured in a 1:5 soil water suspension, and converted to EC_e values if necessary. A table for the conversion is given in appendix 12. Soils with an $EC_e > 4$ dS/m are classed as *saline*.

The bulk *EC* of soil in situ is measured using instruments such as an EM38 (box 7.2).

Box 7.2   ***Field Measurements of Soil Salinity***

The EM38 instrument works on the principle of electromagnetic induction. A transmitter in the instrument generates "loops" of electrical current in the soil, the magnitude of which is directly proportional to the soil EC in the vicinity of the loop. Secondary electromagnetic fields generated by the current flows are intercepted by a receiver, and the summed signal is transformed into a voltage that is proportional to the depth-weighted EC_{soil}. The effective depth of measurement is approximately 1–2 m, depending on whether the EM38 is placed horizontally or vertically, respectively. An example of an EM38 placed in a vertical position on the soil is shown in figure B7.2.1. The EM38 reading is influenced by soil texture and water content, so the instrument should be calibrated against EC_e measurements (Rhoades and Miyamoto 1990). A depth-weighted value of EC_e, calculated from the EC_{soil} for the EM38 in a horizontal position, can be used as a single index of soil salinity to a 1-m depth.

Figure B7.2.1 An EM38 salinity sensor. Photograph by the author.

7.2.2.3   *Grapevine Tolerance of Salinity*

The main effect of salinity on vines is through the uptake of Na^+ or Cl^- ions. Not only can these ions adversely affect growth, but high concentrations in the grapes are also detrimental to wine quality. The Office International de la Vigne et Vin (OIV) sets 60 mg /L as the upper limit for Na^+ in wine. Grapevines are moderately sensitive to salt as shown by the data in table 7.3. More specifically, Na and Cl concentrations in the soil's saturation extract should be <3 cmols charge/L (corresponding to <690 and <1050 mg/L of Na and Cl, respectively). High concentrations of B are sometimes associated with saline soils—ideally, the B concentration in the saturation extract should be <1 mg/L, and certainly not >3 mg/L.

As indicated in table 7.3, some rootstocks, especially those bred from *V. berlandieri*, are more tolerant of salinity than *V. vinifera* itself. In Australia and South Africa, the rootstocks 140-Ru (Ruggeri), 1103P (Paulsen), and 99-R (Richter), which have a *V. berlandieri* parent, have shown moderate to very good salt tolerance. Also, in the Australian irrigated areas, the rootstock Ramsey is particularly effective at excluding Na from the berries of grafted scions. Toxicity in own-rooted vines is likely at leaf concentrations >0.4% Na and >0.7% Cl.

Table 7.3   ***The Sensitivity of Grapevines to Soil Salinity***

Salinity Description	EC_e (dS/m)	Reduction in Vine Yield (%)	Comments
Nonsaline	<2	<10	Negligible effect
Slightly saline	2–4	10–25	Vines on own roots begin to be affected
Moderately saline	4–8	25–50	Own-rooted vines severely affected, but vines on rootstocks such as Ramsey (Salt Creek), Rupestris du Lot, 99R, and 140Ru are more tolerant
Very saline	8–16	>50	Grapevines cannot be grown successfully
Highly saline	>16	0	Vines die

Source: Compiled from Neja et al. (1978), May (1994), and Cass (1998)

7.2.2.4   *Maintaining a Salt Balance Under Irrigation*

EC_e is a measure of the salt concentration at the soil's saturation percentage, that is, the weight of water required to completely saturate 100 g of sieved, air-dry soil. As soil in the field dries out because of evapotranspiration, the concentration of dissolved salts in the soil water and hence the effective *EC* increase markedly. Although the amount of water held at saturation varies two- to threefold between a sandy soil and a clay, the magnitude of salt concentration increase on drying is similar, amounting to a fourfold increase between saturation and the *PWP*. Salts accumulate over many cycles of irrigation in irrigated vineyards and should be counteracted by regular leaching to maintain a salt balance. The calculation of a *leaching requirement* (*LR*) for salt balance, and how this affects the overall irrigation requirement, is shown in box 7.3.

The *LR* can be met through the regular irrigation schedule. But with watering strategies such as *RDI* or *PRD* that use drip irrigation, there will be little drainage during the irrigation period. Leaching of accumulated salts should then occur during the winter dormancy period, as a result of either rainfall or a deliberate blanket irrigation.

7.2.2.5   *Overall Salinity Management*

Good salinity management must take into account not only the soil salt balance, but also any off-site effects of salt. Unnecessary leaching of salts into drainage waters creates salinity problems for users downstream. Proper irrigation management must therefore focus on the dual objective of minimizing saline return flows to streams and avoiding a build-up of soluble salts in the soil and groundwater immediately below the soil.

In this context, the distinction between *LR* and *LF*, the *actual* fraction of the applied water that passes through the root zone, is important. Ideally *LF* should equal *LR*, but because of variability in the application and infiltration of water on a field scale, the average *LF* is generally greater than *LR*. This means that more water is draining through the root zone than is necessary for salinity control in

---

Box 7.3   *Leaching Requirements for Salinity Control*

A complete *salt balance* in an irrigated soil would take account of inputs from the atmosphere, salts released by weathering, and fertilizer salts, in addition to salts in the irrigation water. On the debit side, there would be salts removed in crop products, those lost in drainage, and any insoluble carbonates and sulfates precipitated in the soil. However, the effect of inputs other than through irrigation and outputs other than through drainage tend to balance out, so the salt balance simplifies to

Salt input from irrigation = Salt output in drainage below the root zone   (B7.3.1)

That is,

$$EC_{iw}d_{iw} = EC_{dw}d_{dw} \qquad\qquad\qquad\qquad\qquad (B7.3.2)$$

where the subscripts *iw* and *dw* refer to irrigation and drainage water, respectively, *d* is the volume of water per unit area (mm), and *EC* measures the salt concentration in the water. Rearranging equation B7.3.2 gives

$$\frac{EC_{iw}}{EC_{dw}} = \frac{d_{dw}}{d_{iw}} = Leaching\ requirement\ LR \qquad\qquad (B7.3.3)$$

The *LR* is therefore defined as the fraction of the irrigation water (conductivity EC_{iw}) that must pass through the soil to maintain the *EC* at the bottom of the root zone (EC_{dw}) at a specified value. The critical value for EC_{dw} is set by the vine's tolerance of salinity, expressed in EC_e values (see table 7.3) and is calculated from the approximate expression (Rhoades and Miyamato 1990)

$$Critical\ EC_{dw} = 5EC_e - EC_{iw} \qquad\qquad\qquad (B7.3.4)$$

Thus, if the vine tolerance EC_e is 2.5 dS/m and EC_{iw} of the irrigation water is 1 dS/m, we have

$$LR = \frac{1}{5 \times 2.5 - 1} = 0.09 \qquad\qquad\qquad (B7.3.5)$$

This calculation indicates that about 9% of the irrigation water should drain below the root zone to avoid salinity problems. The total irrigation water requirement is then calculated as

$$d_{iw} = \frac{d_{cw}}{1 - LR} \qquad\qquad\qquad\qquad (B7.3.6)$$

where d_{cw} is the vines' water requirement (mm) as determined by evapotranspiration during the growing season.

---

the soil. Critical aspects of overall salinity management under vineyard irrigation are therefore to

- keep the *LR* as low as possible (<0.05) by using water of low EC_{iw},
- schedule the amount and timing of water application by monitoring soil water status or evapotranspiration, so that the actual *LF* is kept as low as possible ($\cong LR$), and
- use salt-tolerant rootstocks where necessary.

7.2.3   *Sodicity*

When the salt concentration in the soil solution is high, Na^+ ions gradually replace Ca^{2+} and Mg^{2+} ions on the charged surfaces. This is an example of cation exchange (section 4.6.1). The *exchangeable* Na^+ percentage (*ESP*) on the charged surfaces is defined as

$$ESP = \frac{Na^+}{(Ca^{2+} + Mg^{2+} + K^+ + Na^+)} \times 100 \qquad (7.2)$$

where the ion concentrations are expressed in cmols charge (+) per kg soil. As the *ESP* rises, the soil may become *sodic*. The critical *ESP* value to define sodicity ranges from ≥ 6 in Australia to ≥ 15 in the United States. The lower *ESP*

Box 7.4   *Cation Exchange and the Gapon Equation*

As irrigation water is concentrated by evaporation, Na^+ ions gradually replace Ca^{2+} and Mg^{2+} on the clays. The change in the proportions of cations on the clays is gradual because, for a given volume of soil, there is a much larger quantity of exchangeable cations than cations in solution. Take, for example, the exchange reaction

$$(Na^+) + (Ca\text{-}clay) \longleftrightarrow (Na\text{-}clay) + {}^1/_2(Ca^{2+}) \qquad (B7.4.1)$$

The equilibrium constant (K_{ex}) for this reaction can be written as

$$K_{ex} = \frac{(Na\text{--}clay)(Ca^{2+})^{1/2}}{(Ca\text{--}clay)(Na^+)} \qquad (B7.4.2)$$

where the ion activities in solution are in molarities. When rearranged, this expression is identical to an empirical equation—the *Gapon equation*—which was derived from many measurements of cation exchange in soils:

$$\frac{[Na\text{--}clay]}{[Ca\text{--}clay]} = K_{Na,Ca} \frac{[Na^+]}{[Ca^{2+}/2]^{1/2}} \qquad (B7.4.3)$$

In this equation, the ion concentrations are expressed in cmols charge (+) per kg clay and mmols charge (+)/L in solution. The Gapon coefficient $K_{Na,Ca}$ is approximately constant, provided the concentration of Na^+ on the clay is small relative to that of Ca^{2+}. Because Ca^{2+} and Mg^{2+} have similar adsorption affinities on most clays, which differ from the affinity for Na^+, equation B7.4.3 can be written as

$$\frac{[Na\text{--}clay]}{[Ca,Mg\text{--}clay]} = K_{Na-Ca,Mg} \frac{[Na^+]}{\left[\dfrac{Ca^{2+} + Mg^{2+}}{2}\right]^{1/2}} \qquad (B7.4.4)$$

From this equation, we know that the ratio of exchangeable Na^+ to exchangeable Ca^{2+} and Mg^{2+} ions will be governed by the ratio in solution of

$$\frac{[Na^+]}{\left[\dfrac{Ca^{2+} + Mg^{2+}}{2}\right]^{1/2}}$$

which defines the *sodium adsorption ratio* (*SAR*).

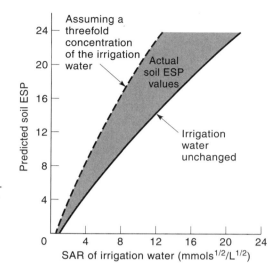

Figure 7.4 The relationship between soil *ESP* and the *SAR* of irrigation water (White 1997). Reproduced with permission of Blackwell Science Ltd.

threshold for Australian soils is attributed to generally low salt concentrations in the soil solutions and the lack of weatherable minerals that can maintain the salt concentration during leaching. An *ESP* ≥ 15 and $EC_e \geq 4$ dS/m define a *saline-sodic* soil in the United States.

Sodic soils present management problems because of the tendency for their aggregates to slake readily when wet and for the clay to disperse (section 3.2.3). This tendency is worsened when pressure is applied, such as from the wheels of heavy machinery in the vine inter-rows, as discussed in section 7.1.2. Many of the duplex soils in Australian vineyards have sodic subsoils (see fig. 3.2e). Because of these potential sodicity problems, the ratio of Na to (Ca + Mg) in irrigation water is important. The influence of this ratio, the *sodium adsorption ratio* (*SAR*), on soil cation exchange is explained in box 7.4.

Results from the western United States show an approximate 1:1 relationship between the *SAR* of irrigation waters and a soil's *ESP*, expressed in cmols charge (+)/kg soil, as shown in figure 7.4. Thus, in the United States and many other countries, an $SAR \cong 15$ mmols charge$^{1/2}$/L$^{1/2}$ measured in the saturation extract is the accepted criterion of a *sodic soil*. In Australia, when the *SAR* is measured in a 1:5 soil to solution ratio, the predicted *ESP* is approximately twice the soil's *SAR*. However, problems associated with a high *ESP* are moderated by a high salt concentration. A saline-sodic soil, for example, can have a stable structure and reasonable permeability, provided that the high salt concentration is maintained. Figure 7.5 shows how the critical *SAR* for decreased soil permeability and unstable structure depends on the salt concentration of the leaching water. Because the *ESP–SAR–*salinity response varies between soils, the relationship in figure 7.5 should be used as a guide only.

The most effective remedy for sodic soil problems is to apply gypsum ($CaSO_4.2H_2O$), which dissolves slowly to provide Ca^{2+} ions. Gypsum should be applied at vineyard establishment along the rip lines, so that it can be washed into the subsoil. However, gypsum can also be applied to established vineyards along the vine row, at rates of 1 t per 100 m of row. The quality of gypsum is based

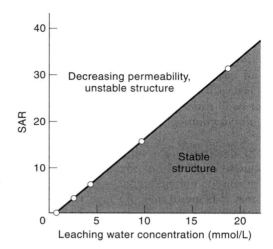

Figure 7.5   The relationship between the critical SAR for soil structural stability and the concentration of the irrigation water (White 1997).

on its S content (>14% S as SO_4^{2-} is suitable) and its particle size (80% <2 mm diameter).

Another effective way to treat sodic soil problems is to grow a permanent grass-based cover crop. This crop helps build up the soil organic matter, encourages earthworms, and generally improves the soil's structure.

7.2.4   *Waterlogging and Drainage*

Problems associated with soil waterlogging are discussed in section 5.6. Transient waterlogging may occur in duplex soils as a result of saturation of the soil above a relatively impermeable B horizon, especially during wet periods in winter. A more serious problem for vineyards occurs when groundwater is present over a large area and the water table occurs near the vine root zone. A predominance of winter rain, together with low evaporation rates, results in excess water draining to the groundwater. In soils of low subsoil permeability, the water drains slowly and the soil may remain saturated for some time. Even soils of higher permeability may become waterlogged because the excess drainage causes the water table to rise. If the water table comes within 1–2 m of the soil surface, water rises by capillary action to the surface, as described in box 7.5. High water tables can also occur in irrigated areas where the application of irrigation water is not properly managed (section 6.6.2).

The Carneros region at the southern end of the Napa Valley in California has heavy clay soils that are poorly drained and tend to waterlog during the winter. Many of the Red Brown Earth soils in the Griffith irrigation area of New South Wales have poor subsoil drainage and water tables that must be controlled. The condition of *restricted spring growth* (*RSG*), which is evidenced by poor shoot growth and die-back in the spring, is more common on heavy textured soils in low-lying sites with poor drainage. Outbreaks of *RSG* have occurred irregularly in Australia since 1940 and also in California, especially after cold, dry winters when the vines have not been irrigated after harvest.

The preferred method of controlling the water table depends mainly on the soil's texture, as discussed subsequently.

Box 7.5   ***The Water Table and Capillary Rise***

Soil water above the water table is drawn upward in continuous pores in the same way that water is drawn up a capillary tube inserted into free water (see fig. 6.1b). This rise of groundwater, called *capillary rise*, depends very much on the soil's pore-size distribution, and is usually not greater than 1 m in sandy soils but can be up to 2 m in some silt loams. A water table within 1–2 m of the soil surface can lead to high soil evaporation losses. As the water evaporates, salts from the groundwater accumulate at the surface; this encourages capillary rise because the gradient in ψ_s adds to the ψ_m gradient (section 6.1.1). This can result in a significant potential gradient upward even in moist soils, when the K value is high and appreciable water movement can therefore occur. If groundwater is saline, the problem of soil salinity is exacerbated.

7.2.4.1   *Draining Clay Soils and Subsoils of Low* K *Value*

The usual method of draining clay soils is to install continuous perforated PVC pipe at depths of 1.5–2 m. Older systems have short lengths of porous clay pipes called *tiles*, usually 75 mm in diameter, laid end to end. Because water enters the pipe from all sides, the drain's performance is much improved if it is embedded in gravel or crushed stone (size range 5–50 mm) or in well-structured topsoil. Such "permeable fill" filters out small soil particles that may otherwise block the drain and reduces the loss in head caused by the restricted number of entry points for water. The pipes drain into ditches that must be deep enough to allow adequate outfall from the pipes.

Because there is resistance to water flow through the soil to the drains, the water table "mounds" to a height h between pipes, as shown in figure 7.6. There is a trade-off between drain spacing L and drain depth d to ensure that the water table at the midpoint between drains does not rise above a critical depth $(d - h)$. The drain spacing can be determined by using the approximate formula

$$L = 2h\sqrt{K_s/J_w} \tag{7.3}$$

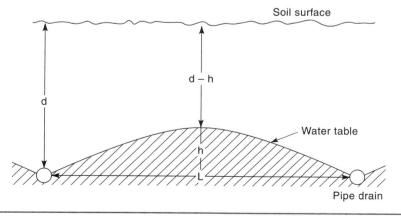

Figure 7.6   Diagram of pipe drains and the "mounding" of the water table between drains.

Box 7.6   *Calculating Optimal Spacings for Different Drain Depths in Clay Soils*

The K_s values for clay subsoils lie between 1 and 5 mm/day. Suppose a pipe drain system is designed to remove 10 mm of surplus water per day (= J_w). Also suppose the height of the water table midway between two drains is not to be <1 m below the soil surface. This means that the maximum ψ_m at the surface, in the absence of evaporation, will be −10 kPa, which should provide an adequate air-filled porosity in the soil. Referring to figure 7.6, we have

$d - h = 1.0$

We can then solve equation 7.3 for different drain depths d to obtain the optimum drain spacing L for soils of known K_s value, as shown in table B7.6.1.

In practice, drains are placed at wider spacings because of the very high cost associated with spacings as close as those listed in table B7.6.1. Deep ripping before planting will increase K_s and hence permit wider, more affordable drain spacings.

Table B7.6.1 *Optimal Drain Spacings for Different Drain Depths*

Soil K_s Value (m/day)	Drain Depth d (m)	h Value (m)	Drain Spacing L (m)
0.001	2	1	0.63
	1.5	0.5	0.32
0.005	2	1	1.4
	1.5	0.5	0.71

---

where K_s is the saturated hydraulic conductivity of the clay layer and J_w is the desired rate of drainage. Box 7.6 provides an example of the use of equation 7.3.

7.2.4.2   *Draining Sands and Gravels with High K Values*

If the hydraulic conductivity of the substrata is high (>1 m/day), but groundwater is a problem, pumps can be used to control the water table. Pipe drains can also be used at much wider spacings than are possible in clay soils. If the groundwater is saline, as in much of the Griffith wine-producing area in Australia, the pumped water cannot be reused but must be transferred to evaporation basins.

7.3   ***Biological Properties***

A biologically healthy soil has the following properties:

- Diverse and active macro- and microorganisms engaged in the decomposition of organic matter, which releases mineral nutrients and provides the residues that improve soil structure
- Earthworms and other macrofauna that burrow through the soil and improve infiltration, drainage, and aeration
- A low incidence of pests and diseases

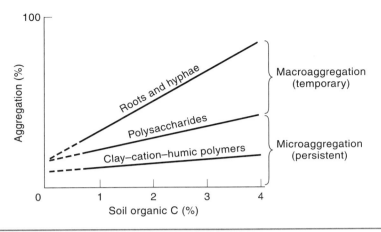

Figure 7.7   The relationship between soil aggregation and the soil's organic C content (redrawn from White 1997).

The role of organic matter, roots, and organisms in stabilizing micro- and macroaggregates in the soil is discussed in section 3.2.2. Key indicators of aggregation are the organic C content and the nature of the organic C components, as shown in figure 7.7. In vineyards, growing cover crops and mulching are two important ways to increase soil organic matter and improve biological activity.

7.3.1   *Cover Crops*

Cover crops are grown in the inter-rows and may comprise annual or perennial species, grasses, legumes, and other broadleaf plants.

Winter annuals. These are grown mainly to protect the soil from erosion during winter and to improve the soil's ability to resist compaction when wet. Sown in autumn to allow germination and growth before the coldest part of the winter, the crop should be mowed in early spring to minimize the risk of frost damage to the vines. The crop is usually incorporated into the soil in early summer to minimize competition for water and nutrients, and to act as a green manure (section 5.3.2). Many species of winter annuals are used, depending on the climate, such as the following:

- Cereals: barley (*Hordeum vulgare*), wheat (*Triticum aestivum*), triticale (a hybrid of wheat and rye), and oats (*Avena sativa*), with and without legumes such as vetch (*Vicia sativa*), field pea (*Pisum sativum*), faba bean (*V. faba*), and subterranean clover (*Trifolium subterraneum*)
- Grasses: fescue (*Vulpia myuros*), annual bluegrass (*Poa annua*), and annual ryegrass (*Lolium rigidum*), with and without legumes

Legume crops fix N_2 and may add a net 50–100 kg N/ha/yr when cultivated into the soil. This extra N may contribute to excess vigor in vines on more fertile soils. Grasses with their fibrous root systems are better for improving soil struc-

Figure 7.8   A vigorous grass cover crop maintained with subsurface drip irrigation in the Lodi District, Central Valley, California. Photograph by the author.

ture, and generally add more organic matter than legumes. If allowed to seed in early summer, a seed bank for subsequent regeneration is built up.

Summer Annuals. These are mainly volunteer grasses, many of which may be native species. Their competition with the vines for water and N can reduce vine vigor on deep fertile soils.

Perennials. These include grasses such as perennial ryegrass (*Lolium perenne*), red fescue (*Festuca rubra*), or native species as in the Central Valley of California; legumes such as white clover (*Trifolium repens*) and strawberry clover (*T. frag- iferum*); and the summer-active, deep-rooted chicory (*Cichorium intybus*). Under irrigation, these cover crops can be very vigorous and difficult to keep out of vine rows, as illustrated in figure 7.8. As is the case with the annual species, these peren- nials compete with the vines for water and nutrients.

Water Conservation. Keeping the rows and inter-rows free of weeds and cover crops decreases competition with the vines and saves water. An herbicide such as glyphosate is commonly used to control weeds in the rows. Clean cultivation of the inter-rows is practiced in many viticultural areas during the summer to min- imize water loss by soil evaporation and transpiration (other than by the vines), but it does adversely affect soil structure (section 7.4.1).

7.3.2   *Mulching*

When a cover crop is mowed in late winter through summer, the mowings can be thrown into the vine row to form a *mulch*. Other forms of mulch are compost, bark chips, and cereal straw (fig. 7.9). Mulches have several beneficial effects in vineyards, as follows:

Figure 7.9   Straw mulch in vine rows, Barossa Valley, South Australia. Photograph by the author.

- Mulch shades the soil, reducing temperatures and soil evaporation during the summer.
- The extra organic matter and moister surface soil encourages biological activity, especially by earthworms.
- Mulches suppress weeds.
- By preventing the breakdown of the soil structure under the impact of rain, mulches enhance water infiltration.

The effect of mulches on soil evaporation is more important when the vines are young and during early summer before the canopy has fully developed, that is, when the *Leaf Area Index* (*LAI*) is low. The *LAI* is defined as

$$LAI = \frac{\text{Canopy leaf area}}{\text{Area of soil surface associated with that canopy}} \tag{7.4}$$

When the *LAI* <1, as much as 70% of the total evapotranspiration is soil evaporation. But at full canopy development (*LAI* >4), soil evaporation accounts for only about 10% of the total evapotranspiration.

As discussed in section 5.7, there is currently a strong trend to make viticulture more sustainable through better use of waste materials and less chemical intervention. To this end, composts made from a variety of materials, including pomace from the winery, are being used as mulches in vineyards. Mulches in the vine rows might encourage the survival of the fungus *Botrytis*, which causes bunch rot in grapes. However, evidence from New Zealand suggests that if a fungus from the genus *Trichoderma* is inoculated into the mulch, *Botrytis* is suppressed. Mulch can also increase the risk of frost damage in susceptible areas (section 3.5.2).

7.3.3   *Control of Pests and Disease*

In addition to beneficial organisms, vineyard soils harbor organisms that can damage or kill the vines. Pest and disease organisms fall into four groups: insects, most seriously phylloxera (*Daktylasphaera vitifoliae*), several types of nematode, vertebrate pests, and pathogenic fungi and bacteria.

7.3.3.1   *Phylloxera*

Phylloxera is an aphid that lives on vine roots and occasionally in galls on vine leaves. The first aboveground symptoms of attack are premature yellowing of the leaves near harvest time. Progressively, the vines decline in vigor, suffer stunted growth, and finally die. The insect is endemic in the eastern United States, where the native species of *Vitis* are resistant. However, European species of *Vitis* are not resistant. When the pest was introduced to Europe around 1860, and subsequently to other parts of the world in the latter part of the nineteenth century, the wine industry based on *V. vinifera* varieties was devasted. A breeding program was established in France to develop hybrids of the favored European varieties as scions, grafted onto resistant American rootstocks. This ongoing program has been so successful that grafted rootstocks should be used as planting material in any area where phylloxera occurs or has occurred in the past.

 Epidemiology. The life cycle of phylloxera is illustrated in figure 7.10. Infected roots swell and become clublike (fig. 7.11), which stops rootlet growth and impairs water and nutrient uptake. The severity of the attack depends on the variety (whether a resistant rootstock or "own roots"), vine age and vigor, and soil type. Phylloxera is a greater problem in heavy clay soils, especially if they are prone

Figure 7.10   The life cycle of phylloxera in California (from Flaherty et al. 1992). Reproduced with permission of Division of Agriculture and Natural Resources, University of California.

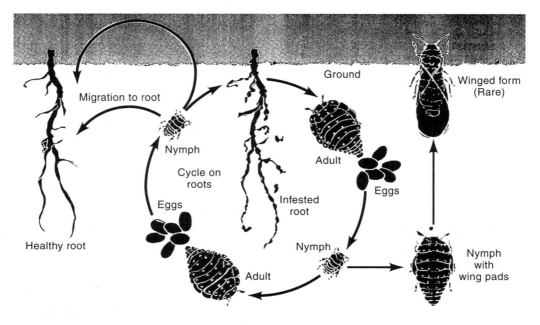

(a)

(b)

(c)

Figure 5.1 Visual symptoms of nutrient deficiency in vines. (a) P deficiency: leaves discolored by reddish-brown blotches in the interveinal area. (b) K deficiency: leaves normal green except for pronounced necrosis extending in from the margins of older leaves. (c) Fe deficiency: leaves overall pale green with extensive interveinal chlorosis. (d) Zn deficiency: leaves dark green except for sharply defined interveinal chlorosis; some shoot stunting. Photographs courtesy of Scholefield Robinson Horticultural Services, Netherby, South Australia.

(d)

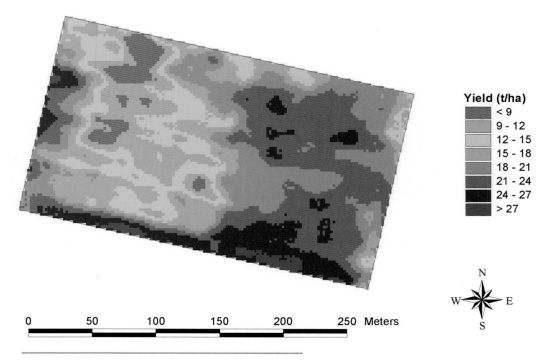

Yield (t/ha)

< 9
9 - 12
12 - 15
15 - 18
18 - 21
21 - 24
24 - 27
> 27

0   50   100   150   200   250 Meters

N
W E
S

Figure 5.5   Grape yield map for precision viticulture
(courtesy of R. G. V. Bramley)

Figure 5.10   Composting of garden waste in windrows.
Photograph by the author.

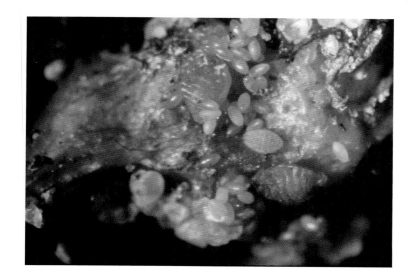

Figure 7.11   Grape phylloxera and eggs on a root of *Vitis vinifera*. © The State of Victoria, Department of Natural Resources and Environment, 2001. Reproduced by permission of the Department of Natural Resources and Environment and therefore is not an official copy. Photograph by Greg Buchanan.

Figure 7.12   Soil profile with compacted layers and poor root development in the Hérault district, Languedoc-Roussillon, France. Photograph by the author.

Figure 7.14   Soil erosion by water on the lower slopes of the Côte
d'Or, Burgundy. Photograph by the author.

Figure 7.16   Terraced vineyards at Pine Ridge in the Napa Valley,
California. Photograph by the author.

Figure B9.1.1   Cabernet Sauvignon vines showing excess vigor and overshaded fruit on a fertile, clay soil at Rutherglen, Australia. Photograph by the author.

Figure B9.1.2   Pinot Noir vines managed to give an optimum ratio of photosynthetic leaf area to fruit, Carneros, Napa Valley. Bunches in the middle and on the right-hand side of the photograph show optimum shading. Photograph by the author.

Figure 9.2   Grand and Premier Cru Classé vines on the slopes of the molasse du Fronsadais, under the escarpment of the Calcaire à Astéries plateau, St. Emilion. Photograph by the author.

Figure 9.6   A Calcic Humifère Brunisol (Humic Calcareous Brown Soil) under forest in the Côte d'Or. Photograph by the author.

Figure 9.12   Profile of a Red Cambrian Soil (Krasnozem) under vines at Heathcote, central Victoria (scale 15 cm). Photograph by the author.

Figure 9.13   Very old Shiraz vines on a sandy ridge at Chateau Tahbilk, Nagamie Lakes Sub-region, central Victoria. Photograph by the author.

These SHIRAZ Vines were Planted in 1860 and are the Oldest in the World.

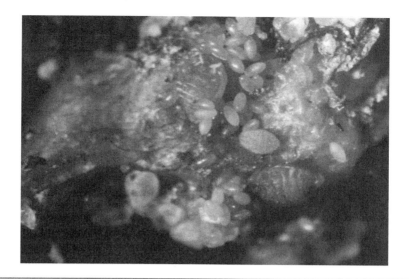

Figure 7.11   Grape phylloxera and eggs on a root of *Vitis vinifera*. © The State of Victoria, Department of Natural Resources and Environment, 2001. Reproduced by permission of the Department of Natural Resources and Environment and therefore is not an official copy. Photograph by Greg Buchanan. See color insert.

to waterlogging, than in sandy soils. In Australia, phylloxera is not expected to invade soils with >70% sand. For example, in the Nagambie Lakes Subregion in central Victoria, a small plot of Shiraz vines survived phylloxera at the end of the nineteenth century because they were growing on a sandy ridge in an alluvial plain of clay loams (section 9.7.3).

Because phylloxera starts from a point source and spreads outward in a circle, the organisms are most easily found on mildly affected vines at the edge of an infestation, for example, on fibrous roots within 0.5 m of the base of the vine during summer (phylloxera are dormant in winter and shelter under bark on the roots). As the population increases, new areas of infestation occur at other sites around the original. The spread is rapid in the first 3 to 4 years, and after 8 years or so the vines are so affected as to be unproductive and subsequently die. The organism spreads on planting material and machinery moved from an infested vineyard to a clean one. Dispersal of the winged form (see fig. 7.10) is uncertain, but flood or furrow irrigation water can move the pests, as can soil erosion by water.

Control of Phylloxera. No insecticide is effective on phylloxera in the vineyard. The prime method of long-term control is to use cultivars that are grafted onto resistant rootstocks. In planting material (cuttings and rooted cuttings or "rootlings"), phylloxera can be controlled by using a hot water treatment, which is also used for the control of nematodes, crown gall, and phytoplasma diseases (appendix 13).

The choice of rootstock is complicated by the need to consider not only resistance to phylloxera, but also resistance to nematodes (section 7.3.3.2), tolerance of drought and salinity (section 7.2.2.3), sensitivity to lime-induced chlorosis (section 5.5.2), and compatability between scion and stock, particularly as it affects vine vigor and grape quality.

Pongracz (1983) claimed that rootstocks with a *V. vinifera* parent would not exhibit long-term resistance to phylloxera. This view was vindicated by the emergence in California, in the early 1980s, of a more virulent phylloxera strain (type B), which overcame the resistance of established rootstocks with *V. vinifera* parents (e.g., AXR#1), necessitating a complete reappraisal of commercial rootstocks. The most favored rootstocks for imparting phylloxera resistance are derived from North American species, the more important being *V. riparia, V. rupestris, V. berlandieri,* and *V. champini*. A summary of the more widely used rootstocks for phylloxera resistance, with their parentage and important attributes, is given in table 7.4.

Table 7.4   ***Commercial Rootstocks and Their Resistance to Phylloxera***[a]

Rootstock	Parentage	Resistance to Phylloxera	Distribution and Other Attributes
Riparia Gloire de Montpellier	*V. riparia*	Very high	Widely used in France, Italy, and California; low vigor; being replaced by *riparia* × *berlandieri* hybrids
Rupestris du Lot (Rupestris St. George)	*V. rupestris*	High to very high	Widely used in nonirrigated soils in California; deep rooted and of high vigor
110 R (Richter)	*V. rupestris* × *V. berlandieri*	High	Widely used in Europe; suitable for droughty sites, but too vigorous for fertile, deep soils
1103 P (Paulsen)	*V. rupestris* × *V. berlandieri*	High	Selected in southern Italy; more drought tolerant than 110 R; vigorous
140 Ru (Ruggeri)	*V. rupestris* × *V. berlandieri*	High	Selected in Sicily, similar to 1103 P; also used in California and Australia
"True" SO-4	*V. berlandieri* × *V. riparia*	High	Widely used in Europe and California, also Australia; low vigor (confused with Teleki 5C in California for many years)
5 C Teleki	*V. berlandieri* × *V. riparia*	High	Used in Australia, California, and parts of Europe; best on clay loam and clay soils
Schwarzmann	*V. riparia* × *V. rupestris*	High	Used in Australia, California and New Zealand; moderate vigor; does not tolerate summer drought
3309 C (Couderc)	*V. riparia* × *V. rupestris*	Moderate	Used in Europe, California, and to some extent in Australia; not suitable for dryland sites
Ramsey (also called Salt Creek)	*V. champini*	Moderate	Widely used in irrigated areas on sandy soils in Australia; very vigorous; similar to Dogridge used in California

[a]Classification of resistance in France is on a scale of 1–5, with class 1 being "immune" and class 5 being "very susceptible." The system in Australia is similar with 4 classes, excluding the "immune" class.
Source: Compiled from May (1994), Wolpert et al. (1994), and Walker (1999)

7.3.3.2   *Nematodes*

Nematodes are small, nonsegmented roundworms that feed on other soil microorganisms or on plant roots. The latter are parasitic nematodes, of which a range of species feed on vine roots, causing malformations and necrosis. Infested vines suffer reduced root function and loss of vigor, without any specific aboveground symptoms. Nematode damage is more severe in sandy soils. Overall, nematodes are a more serious problem than phylloxera in Australian vineyards. The most common nematodes in vineyard soils are outlined here.

Root-knot Nematode (*Meloidogyne* species). These nematodes have a very broad host plant range and are widespread in Australia, especially in the sandy soils of Sunraysia and the Riverland. At least four species—*M. incognito. M. javanica*, *M. arenaria*, and *M. hapla*—are serious pests of grapevines in different regions of the world. The organism is endoparasitic, with the juveniles hatching from eggs in the soil and invading just behind the root tip. The cells of the root cortex swell to form a gall, typically ca. 3 mm in diameter, in which the females breed and lay eggs. A single gall with a succession of females may survive for several years. Cover crop species act as alternate hosts for these nematodes.

Root-lesion Nematode (*Pratylenchus* species). Many species of *Pratylenchus* are found in vineyards, but the most common is *P. vulnus*. The organism is endoparasitic: It does not produce galls on the roots, but migrates through the tissues and impairs vine growth. The lesions formed can predispose the roots to invasion by fungal pathogens such as *Phytophthora* (section 7.3.3.4).

Dagger Nematode (*Xiphinema* species). These organisms are large and ectoparasitic. They migrate through the soil to feed on vine roots, causing swelling of the root tips. The most common species in vineyards worldwide is *X. index*, but *X. americana* also occurs. The former also transmits the Grape Fanleaf Virus (GFLV), which causes problems in vineyards of the North Coast, Napa, and Central Valley of California, and in Northeast Victoria. If a vineyard has both the nematode and the virus, the only effective control is to remove the vines and not replant for 10 years.

Citrus Nematode (*Tylenchulus semipenetrans*). This nematode feeds ectoparasitically on root cells. It is restricted to a few plant genera—importantly, citrus and grapevines—and is common in the Sunraysia and Riverland districts of southeastern Australia where these two crops are commonly grown in close association.

Use of Rootstocks. Resistant rootstocks are the most effective means of nematode control. The species *V. champini* has high resistance to root-knot nematodes, which led to selection of the rootstocks Ramsey and Schwarzmann in Australia. In California, the rootstocks Dogridge, Harmony, and Freedom are favored for their nematode resistance, but the phylloxera resistance of the latter two is questionable.

Muscadinia rotundifolia, a subgenus of *Vitis*, has strong resistance to root-knot and dagger nematodes, which has led to the breeding of the VR hybrids (*vinifera* × *rotundifolia*). However, their use in phylloxera-infested areas is questionable because of their *V. vinifera* parentage. Further details on the range of resistant rootstocks available are given in the review by Nicol et al. (1999).

Apart from the use of resistant rootstocks, nematode damage can be mitigated by good nursery hygiene that ensures cuttings and rootlets are nematode-free (see appendix 13), and any practice that improves root growth and nutrient uptake, for example, the use of manures and fertilizers, or improving soil structure and alleviating soil compaction.

Fumigants such as 1,3-dichloropropene are effective nematicides, especially for nematodes that survive for several years in the roots of previous crops. Because nematodes are found at depths to 90 cm, penetration of the fumigant is a potential problem, so proper land preparation before fumigation is important. This is discussed under Soil Preparation in chapter 8.

7.3.3.3   *Vertebrate Pests*

Rabbits and hares (jackrabbits) eat through the bark of vine trunks and damage the growth cambium. Moles and voles (mainly in Europe), and squirrels and gophers (mainly in North America) attack the roots. These pests are primarily a problem in young vines, which is one of the reasons "vine guards" are used around the trunks (section 7.5.3). The animals can be controlled with traps and baits.

7.3.3.4   *Fungal, Bacterial, and Viral Diseases*

Grapevines suffer from a number of diseases, some of which are soil-borne. For others, soil management can influence the incidence of disease (e.g., mulching and *Botrytis*), or disease control measures can have an impact on the soil. For example, powdery mildew (*Uncinula necator*) is not soil-borne, but the disease is controlled by spraying wettable S when the vines are susceptible. Regular use of S over many years can decrease the topsoil pH (section 5.4.2.2). Powdery mildew persists on vines in winter and develops under favorable temperatures (22–28°C) in summer, especially in crowded canopies.

Botrytis cinerea (bunch rot) is favoured by temperatures of 15–24°C and attacks flowers and berries. It is not soil-borne, but survives on mummified fruits, so that vineyard hygiene is important. However, another canopy disease, downy mildew (*Plasmopara viticola*), overwinters on leaves in the soil. When temperatures are greater than 10°C for at least 24 hr, spores that are splashed onto the foliage by rain (>10 mm), will germinate and infect the leaves. Downy mildew has traditionally been controlled by Cu sprays (Bordeaux Mixture and other Cu-based fungicides), which has led to potentially toxic Cu concentrations in the topsoils of vineyards in the Bordeaux region (section 5.5.2). Another vine disease in California and Europe is the root rot fungus *Armillaria mellea*, which spreads from old oak tree roots. Fungi of the genus *Phytophthora* cause many root diseases, and *P. megasperma* can cause root rot of grapevines, if the vines are waterlogged or where drip irrigation splashes directly onto the trunks of young vines.

Crown gall (*Agrobacterium tumefaciens*) is a bacterium that lives inside the vine and infects wounds in the trunk, forming galls that may kill the vine. It is most severe in cold, moist climates where frosts are common. Although not soil-borne, Pierce's disease (*Xylella fastidiosa*), a serious problem in parts of California, is a bacterium hosted by some grasses and sedges that grow along streams. "Sharpshooter" leaf hoppers pick up the bacteria, spreading them as they feed in adjacent vineyards. Vineyards with permanent cover crops seem to suffer more serious damage (and vine death). Another disease transmitted by leaf hoppers is "grapevine yellows" (flavescence dorée), which is caused by phytoplasma organisms, similar to bacteria, that invade the conducting tissues and may kill young vines.

The Grape Fanleaf Virus (GFLV) and its vector the dagger nematode *X. index* are not lethal but cause yield reductions.

7.4   *Vineyard Soil Management*

A major choice in vineyard soil management is whether to cultivate the soil. There are a number of reasons for cultivation, such as controlling weeds, avoiding frost damage in frost-prone areas, and removing or ameliorating impediments to vine root growth. Clean cultivation of the inter-rows to minimize the risk of frost is discussed in section 3.5.2. Cultivation for weed control and for the removal of impediments to root growth is discussed here.

7.4.1   *Controlling Weeds*

Several options exist for controlling weeds, including complete tillage, strip non-tillage, and complete nontillage.

Complete Tillage. In this practice, a special plow (a French plow) is used in the row early in the growing season, then the soil is hilled in the row and the inter-row fully cultivated with discs and harrows. The practice is common in many French vineyards, in the vineyards of the Napa Valley of California, and in irrigated vineyards of inland Australia. In the Bordeaux and Dordogne regions of France, inter-rows may be alternately cultivated and planted with a cover crop. Complete tillage is common where flood or furrow irrigation is used and also where growers fear herbicide damage to vines. However, such cultivation is expensive, creates dust, damages soil structure, and predisposes the soil to erosion (section 7.5.1). Frequent traffic in the inter-rows leads to soil compaction (section 7.1.2). For example, in parts of Languedoc-Roussillon in France, repeated passes with rotary hoes have led to serious soil compaction, loss of biological activity, poor root development, and unhealthy vines (fig. 7.12).

Strip Nontillage. In this practice, herbicides are applied along the rows, with cultivation and/or mowing in the inter-rows. Preemergence herbicides, such as trifluralin or simazine, are applied in late winter; postemergence contact herbicides, such as glyphosate or paraquat/diquat mixtures, are applied in spring to early summer. Under furrow or flood irrigation, a permanent "berm" (about 1 m wide) is established along each row to prevent irrigation water from leaching herbicides to the vine roots. As well as removing competition for water and nutrients from weeds and cover crops, cultivation helps to control overwintering populations of leafrollers and leafhoppers. Decayed bunches and leaf material are buried and therefore decompose more rapidly.

Complete Nontillage. Preemergence and postemergence herbicides are used in-row, and possibly between rows, although volunteer or sown cover crops are more likely between rows. This practice is most easily implemented under drip irrigation.

7.4.2   *Alleviating Impediments to Root Growth*

Several practices can help alleviate impediments to root growth.

Ridging. Displacement from the inter-rows and ridging of topsoil along the rows may be used when topsoils are shallow over rock and cannot be ripped, the subsoil is dense or impenetrable and cannot be ameliorated, or there is a high water table that cannot be controlled by drainage.

Figure 7.12   Soil profile with compacted layers and poor root development in the Hérault district, Languedoc-Roussillon, France. Photograph by the author. See color insert.

Remedial Tillage to Remove Inter-row Compaction. Moldboard or chisel plowing of the inter-rows loosens compacted soil and, by severing roots, may stimulate new root growth. The best time to do this is in early to midautumn, after harvest. After tillage, wheeled traffic should be minimized. Little can be done for in-row soil impediments to root growth, hence proper site selection and preparation are very important (section 8.4.1).

Deep Ripping. Heavy chisel plows and deep rippers are used to alleviate subsoil compaction and drainage problems. However, this is most effective before planting or at the time of replanting.

Table 7.5 summarizes some quality indicators for vine root growth.

Table 7.5   *Quality Indicators for Vine Root Growth*

"Good" Root Systems	"Poor" Root Systems
Branching system of main roots	Horizontal growth, often contorted
Growth to >50 cm in the soil	Growth is shallow and cluttered
Fine roots (<1–2-mm diameter) found throughout the root zone	Fine roots often lacking or only in the surface soil
Fine roots are white, turgid, and healthy	Fine roots may be brown, shriveled, or dead
All roots are free of disease and pests	Diseases on root surfaces (Phylloxera, ectoparasitic nematodes) or inside the roots (endoparasitic nematodes)

Source: After Cass et al. (1998)

7.5   *Soil and Water Quality Problems*

7.5.1   *Erosion*

The term *erosion* describes the transport of materials, usually from the soil surface, by water and wind. As indicated in chapter 1, past phases of geologic erosion have led to the formation of sedimentary deposits that are the parent materials of many present-day soils. In the short term, accelerated erosion resulting from human activities, especially in agriculture, is of greater concern. *Erosion* and *deposition* are complementary processes because soil removed from the steeper slopes is deposited to a variable extent on flatter, lower slopes and floodplains. Similarly, soil eroded from one area by wind is usually deposited in another part of the landscape (unless blown out to sea).

Erosion is very damaging to soil fertility because mainly the nutrient-rich surface soil is removed, and of that, predominantly the fine and light fractions—clay and organic matter—leaving behind the more inert sand and gravel. The light fractions remain in suspension longer in air or water and so are carried farther. Suspended stream sediment is eventually deposited in lakes, reservoirs, or river estuaries, where it may cause water quality problems.

7.5.2   *Erosion by Water*

Soil loss by water erosion depends on the potential of rain to erode, called rainfall *erosivity*, and the susceptibility of soil to erosion, the soil *erodibility*. We discuss these two factors separately, although they can interact in the field.

7.5.2.1   *Rainfall Erosivity*

Erosivity depends on the physical characteristics of rainfall, primarily the *kinetic energy* (*KE*) available to break down aggregates and dislodge soil particles. Soil particles are detached as part of raindrop splash (fig. 7.13). During splash, the particles are transported a few centimeters. If it occurs on a slope, more of the splash lands downslope than upslope, so there is progressive movement of soil downslope. If water ponds on the surface, lateral movement can occur. Such runoff achieves a greater velocity and hence erosive capacity on steeper slopes. Erosion by water is therefore a two-step process, involving the detachment of soil parti-

Figure 7.13   Raindrop splash at a smooth soil surface (White 1997, courtesy of D. Payne). Reproduced with permission of Blackwell Science Ltd.

Box 7.7   *Calculating the Kinetic Energy of Rain*

The *KE* of rain is given by the equation

$$KE = \frac{1}{2} \times \text{Raindrop mass} \times (\text{Velocity})^2 \qquad \text{(B7.7.1)}$$

Raindrop mass is directly related to drop size. In any one storm, raindrop sizes vary, but the median diameter is 2.5–3 mm for rainfall in temperate regions. The velocity in equation B7.7.1 is the terminal velocity, which increases to a maximum of about 9 m/s for drops with a diameter of 5 mm. Primarily because of the interactive effects of drop size and velocity, the *KE* of rain per mm of rain is found to vary with the rainfall intensity. Although this relationship changes with latitude (temperate versus tropical regions) and with the duration of rain events, the general result is that *KE*/mm increases for rainfall intensities up to ca. 75 mm/hr. The frequency of high-intensity storms (>25 mm/hr) is only ca. 5% for rain in temperate regions, compared to 40% in tropical regions, so temperate rainfall is much less erosive than tropical rainfall.

---

cles by raindrop impact, and the transport of soil particles by raindrop splash and running water.

The energy of falling rain is much greater than the energy of surface runoff generated by that rain. The *KE* of rain depends on its physical characteristics, particularly the rainfall intensity, as discussed in box 7.7.

Indices of rainfall erosivity have been developed on the basis of rainfall *KE* and intensity. One widely used in temperate regions is the *EI index*, calculated as the product of the *KE* of a storm and its maximum 30-minute intensity. The latter is the maximum rainfall during any 30-minute period, expressed in mm/hr. The *EI index* was developed for the United States. Values can be summed over periods of time to give weekly, monthly, or annual values for erosivity. The *EI index* is used in the *Universal Soil Loss Equation* (*USLE*) (see box 7.8).

7.5.2.2   *Soil Erodibility*

Erodibility is the reciprocal of the resistance the soil offers to erosion. It is determined by three broad factors that interact.

Soil Physical, Chemical, and Biological Properties that Affect Aggregation and Aggregate Stability. Properties such the particle size distribution and type of clays, exchangeable cations and organic matter content, and roots and fungi as stabilizing agents mainly affect the detachment process of erosion. Aggregate size and surface roughness operate more on the transport process. For example, if aggregates at the surface slake under raindrop impact, and runoff occurs, the soil particles can then move; some may block pores, reducing the infiltration rate, whereas others may move in the runoff water, possibly initiating saltation (section 1.3.1.2). If aggregates slake *and* disperse (as in sodic soils), the blocking of pores is more severe, and the smaller clay particles are available to move in runoff. The soil surface becomes flatter and smoother, so there is less resistance to runoff.

Topography. The effect of erosive forces (raindrop splash and transport) is greater on sloping land. The effect of slope steepness is confounded with the length

Figure 7.14   Soil erosion by water on the lower slopes of the Côte d'Or, Burgundy. Photograph by the author. See color insert.

of the slope (affecting the volume and velocity of water) and with surface cover (affecting the resistance offered to water flow).

Land Management. This complex factor includes the effect of land use and land management practices. Bare soil is always more susceptible to erosion than soil under vegetation. In the case of vineyards, susceptibility to erosion depends on the time of year (winter versus summer), the distance between rows, and whether the inter-rows are cultivated or under a cover crop. In the Côte d'Or of Burgundy, clean cultivation of inter-rows on sloping land leads to considerable soil erosion during summer thunderstorms (fig. 7.14). In the Willamette Valley of Oregon, vignerons prefer to grow winter cover crops to protect the soil from erosion. Generally, the difference in erosion on the same soil under different management practices is greater than the difference in erosion between different soils under the same management.

7.5.2.3   *Calculating Erosion Rates*

The interaction between factors that determine the rainfall erosivity and soil erodibility at a particular site is shown in figure 7.15. An explanation of the widely used *Universal Soil Loss Equation (USLE)* is given in box 7.8.

7.5.3   *Erosion by Wind*

As with erosion by water, wind erosion depends on the potential of the wind to erode (according to its energy) and the susceptibility of the soil. Wind erosion is a serious problem only in low rainfall areas (<250–300 mm annually), because only dry soil blows. However, a good cover of vegetation protects even dry soil.

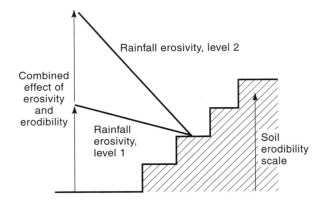

Figure 7.15   The interaction between rainfall erosivity and soil erodibility (redrawn from Hudson 1995).

| Box 7.8 | *Estimating Erosion Loss with the Universal Soil Loss Equation* |

The *USLE* is an empirical model developed from many erosion studies in the United States. The main factors influencing soil loss are multiplied to give an estimate of the average annual soil loss (*A*) from arable land under various cropping systems, according to the equation

$$A = R \times K \times L \times S \times C \times P \qquad \text{(B7.8.1)}$$

In this equation, *R* is a measure of the erosive force of rainfall and runoff, and *K* is the soil erodibility factor. The length factor *L*, slope factor *S*, crop management factor *C*, and conservation practice factor *P* are all ratios of the soil loss from the field under observation relative to the loss from a field under specified standard conditions, namely,

- a length of 22.6 m,
- a slope of 9%,
- cultivated bare fallow, and
- bare soil that is plowed up and down the slope.

Under the standard conditions, the product of *L*, *S*, *C*, and *P* is 1, and the annual soil loss is given by *R* × *K*. For a given site, the land manager has little control over *R* and *K*. Thus, to reduce the average annual soil loss to an acceptable value, the values of *L*, *S*, *C*, and *P* must be reduced.

The *USLE* has been a successful predictive tool in the United States, as well as the stimulus for similar empirical model development in other countries. However, *USLE* has important limitations: it applies only to cropland, and it cannot be used in catchments to predict stream sediment yields and the off-site effects of erosion. With the availability of additional erosion data since its first development, *USLE* has been updated to the *Revised Universal Soil Loss Equation* (*RUSLE*). Additionally, process-based models of erosion applicable to a wide range of scales and land uses are being developed.

Bare soil surfaces are the most susceptible, and so areas where the fetch of the wind is long, that is, where there are no obstacles such as "tree belts" to act as windbreaks.

Vineyards do not normally suffer from wind erosion, except possibly when the vines are young. Windblown soil particles can damage young vines, which should be protected by plastic vine guards around their trunks. In mature vineyards during the summer, the leafy canopies of the vine rows provide an excellent windbreak. During the winter, when the vines are leafless, the soil surface is usually moist so that even clean-cultivated vineyards are not susceptible to wind erosion. If the soil surface is left reasonably rough (because it is not a seed bed), the resistance offered to wind movement is high, and susceptibility to erosion is reduced.

7.5.4   *Control of Soil Erosion in Vineyards*

Erosion control in vineyards involves a combination of managing the soil properties (hence modifying the *K* factor), modifying the slope and length factors (or *L* and *S* combined), and modifying the conservation practice *P*, in some instances in combination with the crop factor *C*.

Erodibility K. All the factors that improve soil structure, particularly as discussed in sections 7.3.1 and 7.3.2 are relevant to making soil intrinsically less susceptible to erosion. The nonwetting sands of Padthaway in South Australia are susceptible to erosion because of their low infiltration capacity. This problem is overcome by mixing subsoil clay with the sandy surface at the rate of 75–100 t/ha. Polyacrylimide wetting agents can also be applied to the surface.

Slope Length L. In Australian vineyards on duplex soils, where the A horizon is shallow or there are drainage problems, the topsoil is formed into ridges along the rows. If the row direction is across the slope, the ridges have the additional benefit of preventing downslope runoff and minimizing erosion. But if the rows run up and down the slope, runoff may be encouraged and the erosion risk increased.

Slope S. In vineyards, the slope factor is most commonly modified by *terracing*. This involves cutting a series of steps into the hillslope to improve infiltration and divert surplus water across the slope rather than down the slope (fig. 7.16). Some of the more famous terraced vineyards, such as those in the Douro Valley of Portugal, were constructed by hand and are supported by stone walls. This process is extremely expensive, but nowadays terraces are made by bulldozers. The practical upper limit of the slope for machine-made terraces is about 20%. Terraces are not normally constructed on slopes <7% because erosion control can be achieved more cheaply by contour banks. In the Napa Valley of California, erosion control measures are required for new vineyards on slopes >5%, because of the extreme environmental sensitivity about runoff from vineyards (normally clean-cultivated), which may carry sediment and chemical residues into streams.

Conservation Practice P. In the Côte d'Or of Burgundy, the most favored vineyard sites are on east-facing slopes of the limestone escarpment. The vine rows run up and down the slope, and erosion is a problem (see fig. 7.14). This is an example, common in viticulture, of a direct clash between preferred growth requirements and the preferred soil conservation practice. However, where vine rows

Figure 7.16   Terraced vineyards at Pine Ridge in the Napa Valley, California. Photograph by the author. See color insert.

are planted across the slope, the erosion can be further reduced by the use of contour ridges or banks. These structures are designed to reduce the length of slope from which runoff is generated, and hold water on the slope longer, so that as much water as possible enters the soil. On low to moderately steep slopes, surplus water from contour banks or terraces can be carried away by a grassed waterway, but stone or concrete-lined drains should be used on the steepest slopes.

Crop Management C. Practices aimed at reducing *C* must focus on not leaving the soil bare when rainfall erosivity is high. The most effective practice is to use in-row mulches and to grow a healthy cover crop, so that the soil surface is protected and soil binding by the plant roots is strong.

7.6   *The Fate of Applied Chemicals, Wastes, and Nutrients*

Increasingly, there are concerns about the off-site impacts of vineyard chemicals, waste from wineries, and nutrient losses, especially of N and P, on surface water and groundwater quality. The key to minimizing off-site impacts is a vineyard management plan based on an informed choice of practices and good record keeping. This is the basis of an *Environmental Management System* (*EMS*), which should cover both the production and environmental aspects of viticulture.

7.6.1   *Pesticides, Fungicides, and Herbicides*

Many synthetic organic chemicals intended to kill a "target" organism are used as pesticides and herbicides, and to a lesser extent as fungicides. The *environmental impact* of such a chemical depends on its effect on nontarget organisms—benefi-

Box 7.9   *Persistence of Chemicals On- and Off-site*

A chemical's persistence at the point of application in a vineyard depends mainly on its

- volatility (hence loss as a vapor) and solubility (hence loss in runoff and drainage water),
- sorption by soil mineral and organic matter, and
- decomposition by chemical (including photochemical) and/or microbiological processes.

A chemical that has mobility and is not easily degraded may become widely disseminated in the environment, as did the organochlorine insecticides such as dieldrin and DDT when heavily used in the 1950s and 1960s. On the other hand, the bipyridyl herbicides paraquat and diquat are immobile and very persistent at the site of application because of strong adsorption by soil clay minerals.

The sorption of many organic pesticides and herbicides is governed by the soil's organic matter content and the affinity of the compound for organic matter, expressed by its K_{OC} value. The smaller the K_{OC} value, the lower the sorption of the compound and the greater its potential mobility. As table B7.9.1 shows, there is an approximate inverse correlation between the water solubility of herbicides used in Australian vineyards and their K_{OC} values.

Because soil microorganisms can adapt to consume almost any added C compound as a substrate for energy and growth, microbial decomposition is a most important process determining pesticide persistence. The process generally follows first-order kinetics (section 2.3.5.1). Persistence is expressed in terms of a chemical's *half-life* (equation 2.2). After one half-life has elapsed, the concentration is reduced to one-half of its initial value, to one-quarter after two half-lives, and so on. The half-lives of nonpersistent chemicals are a few days, whereas those of persistent chemicals are measured in months or even years.

Table B7.9.1 *Properties of Herbicides Used in Australian Vineyards*

Proprietary Name of Herbicide and Chemical Name (In Parentheses)	Solubility in Water (mg/L)	K_{OC} (L/g Soil)
Goal (oxyflurofen)	0.12	14,125
Treflan (trifluralin)	0.22	13,300
Ronstar (oxadiazon)	1.0	4,365
Simazine	6.2	138
Solicam (norflurazin)	28	690

Source: Data from Williams (2000)

cial insects, indigenous flora and fauna, especially mammals and fish—and its *persistence*, that is, how long it remains active in the environment. Factors that determine a pesticide's persistence are discussed in box 7.9.

7.6.2   *Waste Management and Reuse*

The composting of pomace is discussed in section 5.7. More and more wineries are using this approach to recover nutrients from the pressed grapes that would otherwise be lost and to solve a waste management problem. The compost can be used as mulch in the vine rows.

The disposal of wastewater from large wineries must also be properly managed to prevent contamination of water supplies. For example, Riverina Wines in the Griffith Region of New South Wales crushes 40,000 t/yr and produces 64 ML of wastewater. Two hectares of laser-leveled land, with pipe drains at 10-m spacing and 1.2-m depth, are used to "filter" this wastewater, with the drainage being discharged into waterways. After allowing for about 1000 mm/yr of evaporation from the holding lagoon, approximately 85 mm of wastewater has to be applied every 2 weeks, which means that the soil must have a good infiltration capacity.

The application of winery wastewater to land may cause problems because of the relatively high concentration of salts (800–1800 mg/L), high SAR values (3–11 mmols charge$^{1/2}$ per L$^{1/2}$), and low pH (4.5–5). Thus, to achieve sustainable management of wastewater in the long-term, the susceptibility of the soil to salinization, sodicity and pH change should be assessed.

7.6.3   *Nutrients and Water Quality*

Because vines are dormant during the winter and are not absorbing nutrients, NO_3^- is vulnerable to leaching from the soil through winter and early spring. However, the presence of a winter cover crop, especially a cereal or grass, reduces the potential for NO_3^- leaching. The leached NO_3^- may reach surface waters and groundwater that are used for domestic purposes, for which the NO_3^- concentration should not exceed 10 mg N/L (United States and Australia) or 11.3 mg N/L (Europe).

Phosphate ions are not readily leached, but they are carried by soil particles, especially clays, that are washed into surface waters by erosion. Once in the water, P ions are progressively desorbed from the sediment and may raise the P concentration above the acceptable limit of 0.05 mg P/L. The enrichment of streams, dams, and lakes by N and P leads to *eutrophication* of the water body, usually manifest by sporadic "blooms" of Cyanobacteria or blue-green algae (section 2.3.2.1). These algae may release toxins that kill fish and other animal life. When the bloom subsides, decomposition of the large amount of dead organic matter consumes all the dissolved O_2, leading to anaerobic conditions and additional fish deaths. Eutrophication, therefore, disturbs the natural balance of the aquatic ecosystem and creates serious problems for water-treatment plants.

The quantity of P removed by erosion depends on the soil P content and the amount of sediment transported. Quantities range from <0.1 kg P/ha/yr in a well-managed vineyard to 1–5 kg P/ha/yr if erosion is severe (>10 t soil/ha/yr). The quantity of NO_3-N leached depends on the drainage passing below the root zone and on the NO_3^- concentration of that drainage water. Drainage during the summer depends on whether the vineyard is irrigated and how efficiently water is applied and used by the vines. Drainage during the winter depends on the surplus winter rainfall, defined as rainfall minus evapotranspiration (see equation 6.10). Some examples of potential NO_3-N leaching in vineyards are given in table 7.6.

Table 7.6   *Examples of Potential Nitrate Leaching from Vineyards*

Mean NO$_3^-$ Concentration in Drainage Water (mg N/L)	Winter Rainfall over 4 Months (mm)	Winter Evapotranspiration (mm)	Drainage[a] (mm)	Quantity of N Leached[b] (kg N/ha)
10	150	175[c]	0	0
	250	175	75	7.5
	350	175	175	17.5
	150	100[d]	50	5
	250	100	150	15
	350	100	250	25

[a]1 mm = 1 L/m^2

[b]These quantities would increase in proportion to an increase in the drainage NO$_3^-$ concentration.

[c]Vines with a cover crop

[d]Vines without a cover crop

7.7   *Summary Points*

In this chapter, factors that affect the soil quality in vineyards were discussed. The major points follow.

- Good soil quality depends on maintaining a favorable physical, chemical, and biological environment for the growth of vines and beneficial soil organisms.
- Important physical properties are infiltration rate >50 mm/hr; minimal compaction, especially in the subsoil (soil strength <2 MPa at the field capacity); an *AWC* >60 mm/m for a sandy soil and >120 mm/m for a clay loam; and an air capacity >10%.
- The supply of macro- and micronutrients should be adequate, with no element imbalances or toxicities. Vine nutrition can be assessed by *soil testing*, for example, for pH, because of its effect on exchangeable Al^{3+} (damaging to roots), and for the availability of the micronutrients Fe, Mn, Zn, Cu, Mo, and B. *Plant analysis* is most useful for measuring N, P, K, Ca, and Mg.
- Soil salinity should be monitored in inland areas and where irrigation is used. *EC* is a simple measure of the total concentration of salts in the soil or in water. The *salinity hazard* of water is low at an *EC* < 0.25 dS/m (ca. 160 mg/L), but very high at an *EC* > 2.25 dS/m (ca. 1440 mg/L). Irrigation water is concentrated two- to threefold in the soil as a result of evaporation. A *leaching requirement* should be used to maintain a salt balance in soil being irrigated.
- Bulk soil salinity is measured using electromagnetic radiation techniques, as with an EM38. This kind of instrument should be calibrated against *EC* measured on soil extracts, such as the *saturation extract EC$_e$* or the 1:5 soil:water extract (used in Australia). A soil with an *EC$_e$* >4 dS/m is considered *saline*.
- High Na and Cl concentrations in grape juice are undesirable for wine making. *Vitis vinifera* varieties are moderately sensitive to salinity; where salinity is a problem, rootstocks such as Ramsey, Schwarzmann, 140Ru, Rupestris du Lot, and 99R should be used.

■ Sodicity often develops in soils that are, or have been, saline. This condition leads to soil structural problems. A *sodic soil* is one with an *exchangeable Na percentage* (*ESP*) >6 (Australia) or 15 (United States). The soil's *ESP* can be predicted from the *SAR* of irrigation water, which is defined as

$$\frac{[Na^+]}{\left[\dfrac{Ca^{2+} + Mg^{2+}}{2}\right]^{1/2}}$$

■ The application of gypsum ($CaSO_4.2H_2O$) is recommended to ameliorate sodicity.

■ *Waterlogging* is a problem associated with a high water table or an impermeable subsoil. Waterlogging is alleviated in clay soils by installing pipe drains and deep ripping to improve water flow to the drains. In sandy soils, the groundwater can be pumped to the surface.

■ A biologically healthy soil has a rapid turnover of organic residues, with a diverse population of organisms and a water-stable structure. The diverse population of organisms facilitates the breakdown of pesticide and herbicide residues, which minimizes their tendency to move off-site and cause damage to nontarget organisms.

■ Ideally, the root louse *phylloxera* should be absent from a healthy vineyard soil, and the *nematode* population should be low. Where this is not the case, resistant rootstocks, bred from the American *Vitis* species such as *V. riparia*, *V. rupestris*, and *V. berlandieri* (for phylloxera), and *V. champini* and *Muscadinia rotundifolia* hybrids (for nematodes), should be used.

■ *Cover crops* of annual or perennial species are sown in many vineyards. These improve soil structure and trafficability, especially under wet conditions. *Mulches* of cover crop mowings, straw, or compost are used in the vine rows to encourage biological activity and reduce soil summer temperatures and evaporation.

■ Cultivation is used to control weeds, but it tends to destroy soil structure. Cultivation is usually practiced in conjunction with flood or furrow irrigation. On slopes, a bare surface is likely to be eroded by water. Wind erosion is rarely a problem in vineyards.

■ *Erosion* removes the more fertile topsoil and induces undesirable off-site effects. The extent of erosion is determined by the interaction between *rainfall erosivity* and *soil erodibility*. Erosion can be lessened by an improved soil structure, by a change in the length and steepness of slope by terracing or contouring, and by management practices such as planting a cover crop.

■ Dissolved P ions and sorbed P carried on eroded soil particles create the potential for *eutrophication* of receiving surface waters, and consequential algal blooms. Excess NO_3^- in the soil after the grape harvest may be leached during winter, raising the NO_3^- concentration of groundwater above the acceptable limit of 10 (United States and Australia) or 11.3 mg N/L (Europe).

8 Site Selection and Soil Preparation

8.1 Soil Is Important in Site Selection

At the Pine Ridge winery in Napa Valley, California, a sign lists six essential steps in wine production. The first step reads

> Determine the site—prepare the land, terrace the slopes for erosion control, provide drainage and manage soil biodiversity.

Determining the site means gathering comprehensive data on the local climate, topography, and geology, as well as the main soil types and their distribution. Traditionally, site determination was done using the knowledge and experience of individuals. Now it is possible to combine an expert's knowledge with digital data on climate, parent material, topography, and soils in a GIS format to assess the biophysical suitability of land for wine grapes. Viticultural and soil experts together identify the key properties and assign weightings to these properties. An example of an Analytical Hierarchy Process is shown in figure 8.1. In this approach, both objective and subjective data were pooled and evaluated to decide the suitability of land for viticulture in West Gippsland, Victoria. In this region with a relatively uniform, mild climate, soil was given a 70% weighting, and the important soil properties were identified as depth, drainage, sodicity, texture, and pH. But in other areas, with another group of experts, a different set of key properties and weightings may well be identified. For example, a similar approach used in Virginia, in the United States, gave only a 25% weighting to soil and 30% to elevation (which affected temperature, a critical factor governing growth rate and ripening) (Boyer and Wolf 2000).

This kind of approach can be refined to indicate site suitability for a particular variety within a region of given macroclimate. For example, Barbeau et al. (1998) assessed the suitability of sites in the Loire Valley, France, for the cultivar Cabernet Franc, using an index of "precocity." Such an index is related to the

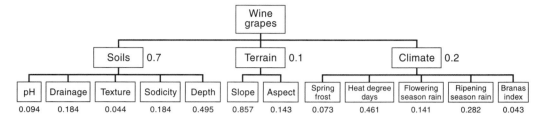

Figure 8.1   An Analytical Hierarchy Process for deciding the biophysical suitability of sites for growing wine grapes. Note that the weights in each group at each level of the hierarchy sum to 1 (redrawn from Itami et al. 2000).

ability of the fruit to accumulate sugar and anthocyanins and to attain a favorable acidity. The *index of precocity for veraison* (*IPV*) is calculated for a site *i* as

$$IPV_i = 100 \left(1 + \frac{V_m - V_i}{V_m} \right) \tag{8.1}$$

where V_m is the mean date (in Julian days) of veraison for all sites studied, and V_i is the veraison date for site *i*. They found that "early" sites (with a desirably high *IPV*) typically were on sandy and gravelly soils that were well drained, whereas "late" sites (low *IPV*) were on clayey and silty soils that had perched water tables in winter. Tesic et al. (2002) obtained similar results for potential Cabernet Sauvignon sites in the Hawke's Bay region of New Zealand (with a uniform, cool maritime climate). These authors developed a site index for the region based on maximum rooting depth, topsoil gravel content, and soil texture (a determinant of soil temperature), as well as the key climatic variables of growing season rainfall and mean air temperatures in October and January.

Although determining the site is only the first step in a chain of events in wine production, it is important because suitable land for vineyards is becoming scarce in many existing wine-producing regions. Also, the investment per hectare in setting up a new vineyard is high, so mistakes made in site selection and preparation can be very costly. The importance of climate and topography in site selection has been discussed by others, such as Gladstones (1992). This chapter discusses techniques for gathering soil data, and how soil and related properties are relevant to site selection and soil preparation for vineyards.

8.2   *Steps in Site Selection*

8.2.1   *Soils Types and Their Distribution*

As pointed out in box 1.3, soil scientists have typically developed general-purpose classifications to group like soils together, creating soil classes or *types*. The output of classification is usually a map showing the distribution of soil classes, with a map legend that describes the central properties of the major classes. However, these maps may not provide information, at an appropriate level of detail, on soil properties important for viticulture. First, there is the question of map scale and second, the question of whether the diagnostic properties used in creating general-purpose classes are relevant to wine grapes.

The relationship between map scale and map accuracy is discussed in box 8.1. Because viticulture is an intensive operation, typically with small, individual vineyards (5–50 ha), soil data should be collected at a high density of observations. With the aid of a Global Positioning System (GPS), the observation sites can be accurately spatially referenced, which facilitates entry of data into the layers of a GIS. In this case, general-purpose soil maps can be replaced with a database of key soil properties, which can be visually displayed through the mapping function of the GIS software. The GIS facilitates the extraction and graphic display of soil properties specifically relevant to viticulture. These data can be used to create a *special-purpose classification* of soils, such as the Australian Viticultural Soil Key developed by Maschmedt et al. (2002). As far as possible, this key uses diagnostic soil properties that are easily recognizable in the field and are described in nontechnical language. The focus is on identifying soil layers that restrict root growth, such as the depth to waterlogging features (mottling and gleying), depth to rock that is soft (rippable) or hard (nonrippable), stoniness, soil structure and consistence, presence and hardness of $CaCO_3$ layers at different depths, soil texture, cracking, and the change in texture with depth.

Although a special-purpose classification may bear little relation to a general-purpose classification such as the Australian Soil Classification (ASC) (Isbell 1996) or the U.S. Soil Taxonomy (Soil Survey Staff 1996), the classes can be approximately allocated to higher level classes of such general-purpose classifications at the regional level. For example, soil types identified in the Australian Viticultural Soil Key have been allocated to Great Soil Group (Stace et al. 1968), ASC, Soil Taxonomy and World Soil Reference Base (FAO 1998) classes. Similarly, in the

---

Box 8.1   *Map Scale and Accuracy*

The accuracy of a soil map is checked by determining whether a chosen soil type, identified at a point on the map, is actually found at that point on the ground. If the density of sampling is too low and soil variability too high, a soil map is unlikely to be accurate.

Normally, soil variability within blocks <100 m^2 (10×10 m) cannot be mapped at an acceptable scale, so this restriction sets an upper limit to the density of sampling. The practical limit is therefore a scale of 1:1,000, for which 1 cm on the map represents 10 m on the ground. This is a *large scale*. Where proximal sensors of soil properties are used (section 8.2.2.1) and digitized data recorded in a GIS, maps of even larger scale can be produced. The scale of a map decreases as distance on the map decreases relative to distance on the ground, so 1:1,000,000 is a *small scale*.

Vineyard soil maps at a large scale ($>$1:10,000) are needed to show local variations as accurately as possible. Many soil maps based on general-purpose classifications are small scale (1:1,000,000 or smaller). Sometimes, to identify soils suitable for vineyards, small-scale maps are expanded to larger scales, for example, by photocopier enlargement or by digitizing the map and entering the data into a GIS. This is a bad practice because the density of the original soil sampling is insufficient to justify showing soil boundaries at the larger scale.

Napa Valley, California, embracing the Napa Valley Viticultural Area (section 9.5.1), soil types separated on the basis of land slope are aggregated into soil series (based mainly on texture and stoniness) that are recognized in Soil Taxonomy.

8.2.2   *Soil Survey*

8.2.2.1   *The Process*

Collecting soil and landscape data for soil classification and mapping is called *soil survey*. In the case of viticulture, the first step in conducting a soil survey is to gather information on *climate* and *geology*. Depending on the vigneron's objectives, such data will identify broad areas potentially suited to the grape varieties, and the style of wine, to be produced. The next step may be gathering remotely sensed data (box 8.2) to delineate differences in vegetation, landforms, and possibly hydrology that could be the outward expression of soil differences. At this stage, the notional soil classes identified will be fairly broad, and further information on specific soil properties, such as salinity, for example, may be sought using an EM survey (see box 7.2). This is called *proximal sensing*. These data can be collected at a high density and mapped on a very large scale.

Once boundaries separating notional soil classes have been marked on aerial photographs, satellite images, or digital maps, soil surveyors must go into the field to check their significance, relocating the boundaries if necessary. This process is called *ground-truthing*.

8.2.2.2   *The Field Survey*

Once an area has been identified as potentially suited to vines, a detailed soil survey is carried out, often by digging pits on a grid pattern, usually at a 75-m × 75-m spacing (about 2.5 pits per ha). This implies that the soil should be mapped

---

Box 8.2	*Remote Sensing for Soil and Land Surveys*

Devices used for remote sensing in soil and land surveys detect electromagnetic radiation (EMR). The sensing may be active or passive. *Active* sensing occurs when the sensing device directs EMR at an object and detects the amount of energy reflected back. Examples are radio wave emissions, radar (wavelength $>10^3$ μm), and laser-imaging radar, and near-infrared, visible, and UV radiation (wavelengths from 10 μm to 10^{-3} μm). *Passive* sensing takes place when the sensing device detects EMR originating from another source, such as γ (gamma) rays emitted from the soil or reflected radiation from the sun (e.g., aerial photos).

Remote sensing relies on detecting differences in the reflected or emitted radiation from different areas of the land over a range of wave lengths. Data are gathered mainly from air- or spacecraft and are recorded photographically or digitally. One example of digital data is a Landsat image from a satellite. Conventional aerial photos must be digitized before undergoing *image analysis* and data processing by computer. In digital format, remotely sensed data provide a natural input to a GIS where they can be put into layers along with other spatial data (e.g., roads, buildings, and streams) and linked to *attribute* data that describe the properties, such as soil properties, of any spatial feature in the GIS.

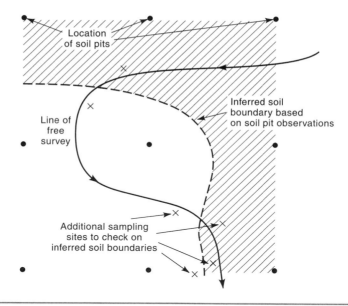

Figure 8.2   Soil class boundaries identified from grid sampling, supported by free survey.

at a scale of 1:7,500, because at that scale 1 cm on the map would represent 75 m in the field (see box 8.1). If the pattern of soil distribution is particularly complex, a 50-m × 50-m grid may be necessary (about 6 pits per ha), justifying a map scale of 1:5,000. With the aid of a soil auger, additional sites can be sampled by conducting a *free survey*, which might be guided by remotely sensed information (e.g., differences in topography or vegetation) or by observable soil features (e.g., a change in soil color, possibly related to drainage, or a stony surface that might indicate a shallow soil profile). The pits or auger holes should be dug to a depth of 1.8–2 m, depending on the depth to a rock barrier. Boundaries between soil types are interpolated from the grid and free survey data (fig. 8.2).

An intensive soil survey, involving remote and proximal sensing supported by detailed ground-truthing, is expensive. Provided that the observation points are accurately located by GPS, all soil and other attribute data can be entered into a GIS, thus facilitating the subsequent application of precision viticulture (section 5.3.5). Grid surveys with a large number of sampling points (>100 pairs) also permit an analysis of any spatially dependent variation in soil properties. Where this is identified, techniques of *geostatistics* can be used to obtain the best estimates of soil properties within localized areas. Specialist texts, such as Webster and Oliver (2000), give full details on geostatistical analysis.

The more detailed the knowledge of soil variation, the better equipped the vigneron is to choose the location of specific cultivars, the design of an irrigation system, appropriate soil amelioration (deep ripping, fertilizers, amendments), vine spacing, and trellising (section 8.4.2).

8.2.2.3   *Describing the Soil Profile*

The basic features of a soil profile—the A, B, and C horizons—and of the main soil-forming processes are outlined in chapter 1. However, to select a vineyard

Table 8.1   *Soil Profile Features Described in the Field*

Soil Feature	Methodology	Reference
Layers or horizons	Thickness measured in cm	
Depth for root growth and thickness of topsoil	Measured to underlying rock or a strongly impeding layer; topsoil is essentially the A horizon	
Texture	Texturing by hand (section 2.2.1.2); confirmed by particle-size analysis (box 2.1)	McDonald et al. (1990)
Color (including mottling)	Munsell Color Chart (a standard color description)	McDonald et al. (1990)
pH	Universal pH indicator (accurate to $+/-0.5$ units)	Raupach and Tucker (1959)
Presence of CaCO$_3$	Add drops of M HCl (vigor of the reaction related to the CaCO$_3$%)	
Soil structure	Estimate the type, size, and grade of aggregates (section 3.2.1)	McDonald et al. (1990)
Consistence	Related to the grade of aggregates at a specified water content (box 2.2)	
Coarse fragments (gravel and stones), including rock type	Estimate according to size ($>$2-mm diameter) and percentage by volume; for rock types, see section 1.3.1.1	McDonald et al. (1990)

Source: Adapted from Wetherby (2000)

site, a description of the soil profile in strictly pedological terms, according to soil survey manuals (section 1.2.2), is unwarranted. In Australia, for example, descriptions of soil profiles focus on the properties that are particularly important for grapevines and for individual varieties or rootstocks within the range available. Emphasis is on identifying potential limitations to production before the vineyard is established. Two popular systems are the *Wetherby system* (Wetherby 2000) and the *SOILpak-PLM system* (McKenzie 2000).

The key features of the Wetherby system are summarized in table 8.1. Simple field tests for salinity (section 7.2.2.2) and aggregate stability (and dispersion) that result from the influence of exchangeable Na$^+$ ions (box 3.2) can also be carried out. Vine performance is inferred from the following properties:

- *Soil depth:* influences root depth and, along with texture, the readily available water *RAW* and deficit available water *DAW* (table 6.3); also affects the plant available water (*PAW*) (table 7.2).
- *Texture:* gives an approximate indication of *RAW*, *DAW*, and *AWC* per unit depth of soil. Texture is also used to convert *EC* measured at a 1:5 soil: water ratio to *EC$_e$* values (appendix 12).

- *Color and mottling:* in the B horizon, orange and red colors indicate good drainage, whereas grey colors indicate poor drainage. Mottles of red in a grey matrix indicate periodic waterlogging.
- *Aggregate consistence:* a strong to rigid consistence, especially when the soil is moist, is likely to indicate impedance of root growth; this may need confirmation by penetrometer tests.
- *Aggregate stability and dispersion:* slaking in water indicates a weak structure; dispersion, as described in box 3.2, indicates varying degrees of sodicity and the likelihood of permeability problems (section 7.2.3).
- *Salinity:* grapevines, especially those on "own roots," are moderately sensitive to salinity. Soils with EC_e values >2 dS/m should be avoided, although vines on Ramsey rootstock can tolerate EC_e up to 4 dS/m (table 7.3).

The SOILpak-PLM (Precision Land Monitoring) system has been adapted for vines from a system developed for cotton growers (McKenzie 1998). It is based on Wetherby's system, with some modifications, for example, sampling at fixed depths of 0–30, 30–60, and 60–90 cm down the profile. This approach gives three layers of data, amenable to GIS entry, which can be mapped consistently across any range of soil variation in the field. Individual Soil Factor Maps (e.g., soil pH) can be produced for each layer and linked to Soil Improvement Maps, as shown in table 8.2. The Soil Improvement Map quantifies the treatment that may be necessary, for example, the amount of lime per ha needed to raise the pH of an acid soil to the desired level and the method of application (section 5.5.3). The information in the Soil Improvement Map can also be converted into a Cost of Repair Map and summed across the GIS map units to give a single cost of development for the site. Additionally, Soil Factor Maps can be used to create a map of on-site and off-site environmental risks, such as erosion, nitrate leaching, soil compaction, and subsoil acidification. This information can be the basis for an Environmental Management System.

Table 8.2   **Soil Factor Maps and Related Soil Improvement Maps Derived from the SOILpak-PLM System**

Soil Factor Map	Soil Improvement Map
pH	Rates of liming material (kg/ha) to raise the soil pH to a desired level
Severity of compaction (based on aggregate consistence or penetrometer test of soil strength)	Deep tillage or ripping recommendations
Dispersion in water (visual assessment)	Gypsum recommendations
Depth to mottled or gleyed subsoil	Recommendations for subsoil drainage or for "ridging" to increase topsoil depth
RAW (based on texture and soil depth)	Guidelines for the design of an irrigation system
Nutrients	Type and rate of fertilizer to be applied

Source: Adapted from McKenzie (2000)

8.3   *Related Factors in Site Selection*

8.3.1   *Biological Factors*

Soil physical and chemical tests should be complemented by tests for disease and pests, such as nematodes. Knowledge of the previous cropping history is important. Land that has been used for vegetable and potato growing, for example, often has a high nematode population. Samples of moist soil can be assessed for the number and type of nematodes present, and to determine whether fumigation is necessary (section 7.3.3.2). Land with old oak trees may be infected with the oak root fungus, *Armillaria mellea*. Replanted vineyards on land previously under vines may have a phylloxera problem. The GFL virus is soil-borne and transmitted by the dagger nematode.

One important decision facing the vigneron is whether to grow own-rooted vines or grafted vines. Cuttings of the former are cheaper, but rootstocks give protection where there is a danger of phylloxera and/or nematode attack, as discussed in sections 7.3.3.1 and 7.3.3.2, respectively. Certain rootstocks are also more salt-tolerant (section 7.2.2.3) than own-rooted vines. However, most of the rootstocks derived from American species of *Vitis* are more susceptible to Fe-chlorosis on calcareous soils than *V. vinifera* (section 5.5.2).

8.3.2   *Quantity and Quality of Water*

A reliable supply of high-quality water is vital for vineyards that need irrigation. Different countries have different laws governing the rights to underground and surface waters—the amount extracted in any one season and the degree of security of the water entitlement. The total quantity of water required is set by that needed in a dry year, as shown in box 8.3. The rate of water supply should be

---

Box 8.3	*Estimating Water Needs for an Irrigated Vineyard*

In the Central Valley of California, which has hot, dry summers, nearly all the vineyards are irrigated. Peak demand occurs in July, when the average daily water need is about 0.33 acre-in. (Verdegaal 1999). The following calculations, in metric units, illustrate the quantity of water required. Conversion from metric to U.S. units is explained in appendix 15.

Thus, 0.33 acre-in. of water per day is equivalent to 34 m^3, which, if spread over 1 ha of soil, amounts to 3.4 mm of irrigation water per day. If irrigation were scheduled every 7 days, the quantity of water required per week would be 0.238 ML per ha.

This figure could be adjusted down or up, depending on whether some form of *RDI* or *PRD* was practiced or cover crops were grown in the inter-rows.

The inland vineyard areas in Australia also have high water requirements for irrigation. For example, in the Griffith irrigation area, the average application for drip-irrigated vines in midsummer is about 18 L per dripper per day. If there are 2,500 drippers per ha, this amounts to 4.5 L/m^2 (4.5 mm) or 0.045 ML/ha/day of water. The quantity required per week in the Griffith irrigation area, 0.315 ML, is higher than that in the Central Valley, California.

sufficient to supply vine needs in midsummer (around veraison). Where overhead sprinklers are used for frost avoidance in the vineyard, additional water may be necessary.

Whether obtained from surface storage or underground, the water should be analyzed for the following:

- pH
- Total salts (by *EC*) (section 7.2.2.1)
- Na, Cl, and B concentrations, for potential toxicities and effects on wine quality (section 7.2.2.3)
- Ca, Mg, and carbonate concentrations for assessing water "hardness" and for calculating the *SAR* of the water (section 7.2.3)
- Fe concentration, because precipitates of $Fe(OH)_3$ can block drippers and even pipe lines (section 2.2.4.3)

General criteria for the salinity hazard of irrigation waters are given in section 7.2.2.1. Because irrigation water may be concentrated two- to threefold by evaporation in the soil, water with $EC > 1$ dS/m (640 mg salts/L) should be avoided for vines, except when salt-tolerant rootstocks are used. Drip irrigation permits the use of water with higher *EC* values than water used in spray or flood/furrow irrigation.

Availability of Recycled Water. Viticulture must compete for water with other rural and industrial users. Increasingly, vignerons are seeking to use recycled water, or wastewater from cities, especially for drip irrigation where there is no danger of the water contacting the fruit. For example, the McLaren Vale region of South Australia has restricted use of groundwater, and irrigation with secondary-treated sewage effluent is being introduced for up to 1500 ha of vineyards. The effluent is relatively high in nutrients (approximately 20, 9, and 24 kg/ha of N, P, and K for every 100 mm applied), so fertilizer inputs should be adjusted. Similarly, Scotchman's Hill on the Bellarine Peninsula, Victoria, which has an annual rainfall of approximately 500 mm, now uses "grey water" from a local sewage works for drip irrigation, after water has been stored in holding lagoons for 18 months. The concentration of salts in such wastewater is relatively high (ca. 1400 mg/L), which can create problems after regular use in vineyards unless the soil is leached during the winter to remove excess salts.

8.3.3   *Slope, Aspect, and Presence of Water Bodies*

Soil drainage is generally better on slopes than on flat lands and valley bottoms. Slopes also provide *cold air drainage*, which helps to avoid excessively low temperatures in winter and frost damage in spring. Vines require 200 frost-free days, and frosts following bud burst can result in serious crop losses. Beginning in autumn and throughout winter and spring, the sun's rays pass through a greater thickness of the earth's atmosphere than in midsummer. This effect, which decreases the intensity of sunlight received by the soil and vegetation, is most marked at the highest latitudes where vines are grown (45–50° N or S). At such latitudes, steep slopes facing the equator also absorb more solar radiation than does flat land, because the angle of incidence at the soil surface is closer to 90° (fig. 8.3).

Thus, depending on geology and topography, a predominantly southerly aspect in the Northern Hemisphere and a northerly aspect in the Southern Hemisphere is important in cool climate regions, for example, Burgundy and Cham-

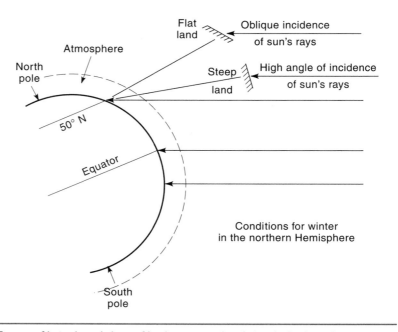

Figure 8.3   The influence of latitude and slope of land on received and absorbed solar radiation.

pagne in France and the Otago Valley in the South Island of New Zealand. Vine rows in Burgundy usually run up and down the slope and are oriented east–west. However, rows up and down the slope in clean-cultivated vineyards are susceptible to erosion (see fig. 7.14). But in the Rhinegau of Germany, another cool climate region, the preferred vineyard layout is for rows to run north–south on a west-facing slope. In this region, morning fogs are common during the critical autumn ripening period, but these usually clear by the afternoon, so a west- to southwest-facing slope maximizes warming of the soil.

In hotter climates, such as southern Spain, the Central Valley of California, and the inland irrigation areas of Australia, minimizing the exposure of fruit to excessive direct radiation, which causes sunburn, is more important than maximizing soil warming. The positioning of rows may influence the susceptibility of vines to fungal diseases in humid regions, because surface drying of foliage and fruit is enhanced by wind turbulence when the rows are perpendicular to the prevailing wind direction (Jackson 2000). Row orientation and the presence of tree belts provide important protection from wind in maritime areas, for example, in many of the viticultural areas of New Zealand and coastal regions of California.

In undulating topography, cold air drains from high points, down valley slopes to valley bottoms, while warm air rises. In cool climates, this cold air drainage makes vines on the lower slopes and valley bottoms susceptible to frost damage at bud burst, and even at flowering. Similar "convection cells" are set up near substantial bodies of water, such as lakes, large rivers, and estuaries, as shown in figure 8.4. The water body is a greater heat sink than the nearby land, so warmer air over the water rises, to be replaced by cooler air draining from the land. Due

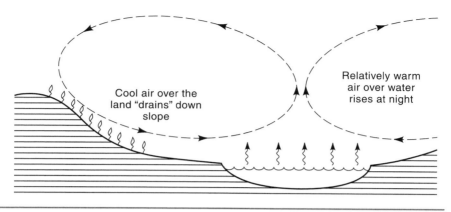

Figure 8.4   A convection cell on a clear night for a vineyard near a large river (redrawn from Gladstones 1992).

to the moderating effect of these convection cells, vineyards near water bodies do not suffer the extremes of temperature, either diurnal or seasonal, that occur in continental climates (section 3.5.3).

8.4   *Vineyard Establishment*

8.4.1   *Soil Preparation*

Soil preparation starts in summer. If the land has been cleared of trees or old vines, all residues must be removed or burnt. The soil should be plowed (usually with a chisel plow) to bring up old roots. *Deep ripping* may be necessary to break up a restrictive layer at depth (an old plow pan, an iron pan, or a concretionary layer; see section 3.2.4). In old vineyard soils, the compaction from traffic in the inter-rows can be serious, especially in sandy soils, but this can be alleviated by ripping. Many duplex soils in Australia have a compacted A2 horizon, which can be improved by mixing with the better structured A1 horizon (Myburgh et al. 1998). On the other hand, vineyards established on calcareous soils in Hérault, France, which have been mixed by deep moldboard plowing, have suffered root growth problems (see fig. 7.12). These soils also have been overcultivated with rotary hoes.

8.4.1.1   *Deep Ripping*

To shatter the restrictive layer, deep ripping is best done during late summer and autumn, ideally when the subsoil is slightly drier than the lower plastic limit, so the bulk soil will not smear as the ripper blade passes through, but rather will shatter. In some cases it may be necessary to sow a winter cereal, to be harvested the following summer, to dry the soil profile sufficiently before deep ripping. The methods and conditions for successful deep ripping are described in box 8.4.

After deep ripping, the land is leveled and prepared for planting. Fumigating for nematodes is most effective when the soil is dry (to maximize the air-filled

Box 8.4   *Methods of Deep Ripping*

Compacted soil layers and pans can be broken up by deep ripping (also called *subsoiling*), preferably to a 1-m depth. Achieving this depth requires a 1.5-m shank and a D8 or similar tractor (fig. B8.4.1). The subsoiler blade and its angle are chosen based on the problem to be alleviated, as shown in table B8.4.1. If the soil is ripped at an appropriate water content and at a 2-m spacing (with wings attached), the fissures created by the blade should intersect at the soil surface. This is important for good drainage.

One direction of ripping is along the intended vine rows. Ameliorants such as lime or gypsum can be broadcast before ripping, which improves their incorporation into the subsoil. The soil is usually ripped in one other direction, and at least one rip line should run downslope to avoid waterlogging in winter. In the Napa Valley of California, three-way ripping with lime and gypsum additions is practiced.

Figure B8.4.1 D8 "Caterpillar" with deep ripping shanks. Photograph by the author.

Table B8.4.1 *Recommendations for Ripping of Subsoils*

Soil Problem	Blade Type	Rake Angle[a]
Hard rock	Do not rip	
Weathered rock	Use wingless blade	
Cemented pans	Use wingless blade	
Clay subsoil	Use winged blade	20° from horizontal
Compacted sands	Use 90° point	

[a]The angle of the blade is critical for proper lift and shattering of the soil, with minimal lateral compaction.
Source: After Cass et al. 1998

pore space for gas diffusion) and temperatures are greater than 12°C. After fumigation, a cover crop is sown to protect the soil during winter, until the vines are planted in spring. Fertilizers can be applied with the cover crop, although if rock phosphates are used, as with lime or gypsum, they should be incorporated along the vine row when the soil is ripped. If the topsoil is shallow, the soil in the vine row may be hilled to increase the depth and improve drainage.

8.4.1.2 *Cover Crops*

The use of cover crops for soil protection and stabilization of structure is referred to in section 7.5.4. The physical and biological basis for the beneficial effect of plant roots, especially grass roots, on soil structure is outlined in section 3.1.1. The species commonly used in cover crops and their management are discussed in section 7.3.1, where their role in controlling vine vigor is also indicated.

8.4.2 *Site Potential and Planting Density*

Preliminary survey data on geology, topography, mesoclimate, and soil are used to evaluate *site potential*, primarily to assess the likely vigor of the vines to be grown. Vigor depends on the interaction among the rootstock and scion, vineyard management, soil, and the environment. Of the soil factors that affect vigor, the main ones are soil depth, *AWC* per m depth, drainage, and soil fertility, primarily N.

Although there are various ways in which these factors can be modified by management (e.g., deep ripping, irrigation, including *RDI* and *PRD*, cover crops, and fertilizer management), the value of these factors at vineyard establishment indicates the site potential (table 8.3). Depending on this potential, the vine spacing can then be chosen to satisfy the dual objectives of *highest quality fruit* from each vine and *optimum yield per ha* of this fruit. Another consideration is that the vineyard should be easy to manage with whatever machinery is necessary.

The general principles for vine spacing, derived largely from research in California, Australia, New Zealand, and South Africa (Smart and Robinson 1991, Weber 1992) are summarized here:

Table 8.3 **Characteristics of Low- and High-Potential Sites**

Low-potential Sites	High-potential Sites
Impervious rock underlying	No impervious rock or other shallow barriers
Shallow soil depth (<0.5 m)	Deep soil (>1.5 m)
Sandy to sandy loam texture, or high proportion of stones (>2-mm diameter)	Clay loam to light clay texture; <5% stones by volume
Low organic matter content (<1% C)	High organic matter content (>2% C), well humified
Weak soil structure, especially in the subsoil	Well-aggregated soil; aggregates stable in water
Poorly drained subsoil (grey gley colors and orange-red mottles)	Well-drained soil throughout the soil profile (uniform orange to red-brown colors in subsoil)

- Low-potential sites need closer vine spacings, and high-potential sites need wider spacing.
- Between-vine (in-row) spacing has a greater effect on yield and quality than between-row spacing.
- Vines in-row should be far enough apart to provide enough cordon space for a balanced fruit load.
- The effects on vine growth of a high potential site can be modified by the type of trellising.

A discussion of the rationale behind these principles follows.

Low-potential sites. Vines will naturally be small because of soil limitations, hence close spacing should be used to increase the production per ha. As planting density increases, the competition between vine roots for water and nutrients increases. If the competition for water is too severe post-veraison, irrigation can be used. Canopies are less dense and more open than on high-potential sites, allowing more light to the fruit and enhanced ripening, especially in cool climate regions.

Vines in-row should be close enough that there are no gaps between them when they are mature, yet they should be far enough apart for an adequate number of buds to be left after pruning to achieve a properly balanced vine. The in-row spacing can be as close as 1 m. Between-row spacing depends on the type of trellising to be used and the size of the machinery available for spraying and other cultural operations. A Vertical Shoot Positioning (VSP) trellis is commonly used on low-potential sites. With over-the-vine equipment, such as that used in France, rows can be as close as 1 m apart. In this case, the vines need to be trimmed to a height no greater than the row spacing to avoid too much mutual shading of the canopies (fig. 8.5). In many Medoc and Côte d'Or vineyards in France, the

Figure 8.5   Vineyard near Bordeaux with close planting and vine trimming. Photograph by the author.

Figure 8.6   Vine row with Lyre trellis and horizontally divided canopy. Photograph by the author.

maximum density of vines is 10,000 per ha (1 m × 1 m). This spacing is about twice the densest spacing used in California and Australia. Although the yield per vine is low, a satisfactory yield per ha can be maintained because of the higher plant density.

High-potential sites. Sites on deep, well-structured loam to clay loam soils with high organic matter content will naturally produce vines of high vigor. Close planting and the application of *RDI* do not necessarily increase root competition and control vigor, because the vines root more deeply and explore a much larger soil volume than they would on low-potential sites. However, yield and quality can be optimized by using row spacings of 3.4–3.6 m and in-row spacings not less than 2 m. The trellising should be a divided-canopy type, such as the Geneva Double Curtain or Lyre (fig. 8.6), or the Scott Henry or Smart Dyson types, all of which allow more cordon length per hectare and an increased crop load commensurate with the vegetative growth. The open canopy also helps to avoid excessive fruit shading. Management to control excess vigor on high-potential sites is discussed further in chapter 9.

8.5   ***Summary Points***

This chapter focused on the proper site selection and soil preparation for growing grapevines. The major topics are summarized as follows:

■ Knowledge of the soil is important for site selection. Soil data relevant to viticulture are used together with climate and topographic data to determine the most suitable sites for growing particular varieties.

■ Generally, soil information derived from a *general-purpose classification* lacks detail on properties specific to viticulture at an appropriate scale. A *special-purpose classification* for viticulture should be created, and, if applicable to a larger area, the classes correlated with a national classification (e.g., Soil Taxonomy in the United States).

- A *soil survey* can provide detailed information on soil variation, usually expressed in the form of a map. Before this is begun, data from remote sensing, geology, and vegetation maps should be examined to identify possible soil types and their distribution. Inferences drawn from aerial photographs or satellite images should be ground-truthed. Digitized data can be entered into a GIS. These data can be augmented with that acquired by proximal sensing at a high density of observations (e.g., a salinity survey by EM methods).

- The soil is best examined in pits dug to depths of 1.8–2 m, preferably on a 50- × 50-m grid (approximately 6 pits per ha). Information on soil variation between pits is obtained by conducting a *free survey*, using an auger. The location of all observation points should be determined by a GPS, so that data can be entered into a GIS. The density of sampling on the ground determines the scale at which the soil is mapped.

- Soil horizons should be identified and the following key properties measured: soil depth, stone content, texture and texture change with depth, color (including any mottling or gleying), aggregate consistence and whether aggregates are stable in water, content and depth of any $CaCO_3$, and salinity (measured by *EC*).

- Detailed knowledge of soil variation can be used to plan the location of specific cultivars, the irrigation layout, and the implementation of precision viticulture.

- A Soil Factor Map (e.g., with soil pH distribution) can be linked to a Soil Improvement Map (with recommendations on liming) and also to a Cost of Repair Map in a GIS.

- Biological factors to be considered are the past cropping history and presence of pests, especially phylloxera and nematodes, and soil-borne diseases. These will influence the choice between own-rooted vines and rootstocks.

- If irrigation is needed, the quantity and quality of water supplies should be assessed. Quantity estimates should be based on peak midsummer demand. The *EC* of irrigation water should generally be <1 dS/m, except where salt-tolerant rootstocks such as Ramsey, 140Ru, or 99R are used. Where natural supplies of water are scarce, recycled water ("grey water") may be considered.

- Land slope and aspect are important site factors because they influence soil properties, the amount of solar radiation received, the choice of row orientation, and temperatures. Proximity to large water bodies moderates the diurnal temperature range and seasonal temperature extremes.

- Preplant soil preparation generally involves deep ripping, to break up layers that may impede roots or drainage, and fertilizing. Fumigation may be carried out if the nematode population is high. This is done during late summer and autumn, and a cover crop is sown to protect the soil during winter until the vines are planted.

- Preliminary survey data on geology, mesoclimate, and soil are used to evaluate *site potential*, and hence the likely vigor of the vines. Depending on this potential, vine spacing (both between rows and in rows) is then chosen to satisfy the dual objectives of *highest quality fruit* from each vine and *optimum yield per ha* of this fruit. Vine vigor is also influenced by factors such as trellis type, canopy management, cover crops, fertilizer, and water supply.

9 *Soil and Wine*

Soil Variability, Terroir, and Scale

The concept of *terroir* as a complex interaction among soil, climate, biology, and human intervention is introduced in section 1.1. The belief that the soil in a particular vineyard imparts a distinctive character to the resulting wine is strong in Europe, but less so in the New World. The special character or personality of a wine may be confined to just one small block, less than 0.5 ha, for example, the "core block" within L'Enclos of Château Latour in the Bordeaux region (Bordelais) of France. Alternatively, a special character may be attributed more widely to wines from an appellation (the commune Pauillac) or to a subregion such as the Haut-Médoc. But soil is very variable in the landscape (chapter 1), so that as the vineyard area increases, the character of a wine is less and less likely to show a distinctive and defining influence of the soil. Soil variation, in combination with a variation in the mesoclimate (section 1.3.2), will mask a clear, intense expression of the underlying *terroir*. The grape variety, cultural practices, and the wine maker will then dominate the wine character.

Thus, the true influence of *terroir* can only be satisfactorily studied for small areas. As pointed out in section 8.2.1, soil information is typically collected at a low sampling density over large areas to produce general-purpose soil classifications. The resulting soil maps are necessarily of a *small scale* (e.g., 1:1,000,000), which means the information about small areas (1–10 ha) is unlikely to be very accurate (see box 8.1). Hence, intensive soil surveys, with at least 6 soil pits per ha, are necessary to study the soil factor in *terroir* when soil variation can be mapped at a *large scale* (1:5,000). Further, with more widespread use of precision viticulture technology, as discussed in section 5.3.5, the variation in specific soil properties (e.g., depth to an impeding B horizon and soil strength) can be measured at intervals of about 2 m and mapped at a very large scale (>1:1,000).

At a small scale (representing a large area), we can make generalizations, such as that soils on limestone or chalk in Burgundy, Champagne, and the Loire Valley in France are highly regarded for producing distinctive wines. French scientists such as Seguin (1986) and Pomerol (1989) attribute this character to the soil's *physical properties*—the structure, the rate of water supply to the roots, and drainage. *Chemical properties* are considered less influential, provided that the nutrient supply is balanced (no surpluses that promote excess vigor or toxicity, and no serious deficiencies). In the New World, Gladstones (1992) argues that rocky and gravelly soils produce the best wines across a wide spectrum of climates. Because soils with stony surfaces are found on the lower slopes of hills (often formed on colluvial material), most of the best sites are on such slopes, rather than on the plains or lowlands. Gladstones suggests the best sites are on slopes with excellent air drainage, on slopes that face the sun directly for part of the day, even in hot areas, are frequently close to water, and preferably on the sides of projecting or isolated hills. These are all factors that tend to minimize temperature variation during the day and maintain higher temperatures at night and in the early morning. Gladstones could have added that rocky and gravelly soils are also often low in nutrients.

However, Halliday (1993) contends that in Australia the prevalence of a summer drought induced a change from the traditional European preferences for sites and soils. Vineyards on sunny slopes of shallow, calcareous soil did not prosper, so vineyards on deep alluvial soils in valley bottoms were preferred, as was the case in the Hunter and Barossa Valleys (also the Napa Valley in California). Now that irrigation, especially drip irrigation, is more widely practiced, vineyards in premium wine-growing regions such as these have expanded upslope. The yields are lower, but the quality is generally higher.

The prime aim in viticulture is to achieve the maximum production of quality fruit per hectare. Depending on the marketing strategy, the vigneron strikes a balance among yield per vine and yield per hectare, the quality of the fruit, and the intensity of berry flavors. The key philosophy in France is that the best fruit, of most intense flavors, comes from vines with a limited yield. For example, Seguin (1986, p. 865) states

> In general, everything happens in the Bordelais as though the temperature and the degree of sunlight enabled the synthesis of a limited quantity of colouring matter, aromas and sapid elements, and it seems that these substances are diluted and deteriorated when there is too sharp an increase in yield.

In many of the best French appellations, the wine yield is limited to 35 hL (ca. 5.8 t/ha). Wine produced from irrigated vineyards is not accepted for the top AOC category. However, many vignerons in the New World, notably Australia, believe that intensely flavored fruit of high quality can be produced from high-yielding vines, provided the vines are "in balance" (Smart and Robinson 1991). The concept of vine balance is discussed in box 9.1.

9.2   *Defining the Terroir*

As indicated previously, *terroir* denotes more than just the relationship between soil and wine character. Wilson's (1998) book *Terroir* was subtitled *The Role of*

Box 9.1   *Vines in Balance*

Balance is achieved when vegetative growth and fruit load are in equilibrium. Chapters 5 and 6 discuss the condition of "excess vigor" that results from too high a supply of soil N and/or soil water during the period of rapid vegetative growth. An example of vines showing excess vigor is shown in figure B9.1.1. The consequences of excess vigor are overshading of the fruit, slower fruit ripening (a particular problem in cool-climate vineyards), less fruit color and flavor, and lower bud fertility the following season. Excess vigor is difficult to control on deep, naturally fertile soils in high rainfall areas. Practices to control vigor include trellising that opens up, or divides, the canopy, shoot trimming in early summer, and leaf removal between fruit set and veraison. These are discussed in detail by Smart and Robinson (1991). Other practices to control vigor involve the choice of rootstock (table 7.4), use of grass cover crops (section 7.3.1), the planting density (section 8.4.2), whether vines are irrigated, and if they are, control of the water supply at critical stages of growth (section 6.5). An example of vines that have been managed to control vigor and achieve vine balance is shown in fig. B9.1.2.

Clearly, management practices interact with the soil and the environment to determine vine balance. To a degree, one factor can substitute for another, such as a restricted soil water supply reducing vegetative growth thus minimizing the need for intensive canopy management. The most important outcome of this interaction is optimum physiological fruit ripening, which is the end result when vines are in balance (Gladstones 1992).

Figure B9.1.1 Cabernet Sauvignon vines showing excess vigor and overshaded fruit on a fertile, clay soil at Rutherglen, Australia. Photograph by the author. See color insert.

(continued)

Box 9.1 *(continued)*

Figure B9.1.2 Pinot Noir vines managed to give an optimum ratio of photosynthetic leaf area to fruit, Carneros, Napa Valley. Bunches in the middle and on the right-hand side of the photograph show optimum shading. Photograph by the author. See color insert.

Geology, Climate, and Culture in the Making of French Wines. Similarly, Pomerol's (1989) survey of French wines and wine regions was in the form of a "geological journey." In section 1.3.1 of this book, the role of *parent material* in soil formation is described. But this book clearly states that soil formation is also influenced by climate, relief (topography), organisms (including humans), and time.

All these factors interact to determine the type of soil at a particular location in the landscape. Hence, in examining the relationship between the soil and wine character, we are inevitably also considering the influence of climate, geology, and human activity, which has received more attention in the past. Human activity in vineyards is especially important in the old viticultural areas of Europe, as revealed by more than 300 years of records of soil building and replacement in Chateau Latour in the Médoc (Johnson 1994) and also on the shallow chalky soils of Champagne (section 1.3.3.2).

Two main approaches to defining *terroir*—geographical and technical—are discussed in the following sections.

9.2.1   *The Geographical Approach*

Traditionally, wine areas in Europe have been delineated strictly on geography, taking account of the geology, soils, topography, mesoclimate, grape varieties, and whether there has been a consistent distinctive character associated with the wines produced over a long time. This approach is the basis of the French AOC system, which came into effect with the establishment of the Institut National des Appellations d'Origine (INAO) in 1935. The French system of categorizing wine is the model for EU wine legislation. The AOC system identifies all the vineyards in particular areas, based on their geographic location, and prescribes the acceptable varieties, viticultural methods, yield, fruit ripeness, and maximum alcoholic strength of the wine. The system offers the consumer a product guarantee through its categories of Appellation d'Origine Contrôlée (AOC), Vin Délimité de Qualité Supérieure (VDSQ), and Vin de Pays. However, the AOC regulations make it difficult for growers, especially those outside the historically favored regions of Burgundy, Bordeaux, and Champagne, to experiment with new varieties and production systems. Also, the relationship between a geographical area and the character of a wine becomes less meaningful for "catchall" appellations, such as Bordeaux and Champagne (Robinson 1999).

Similar, but generally less restrictive, quality recognition systems are in place in Italy, Spain, Portugal, and Greece in Europe, and in Argentina. The United States and Australia also have systems that define viticultural regions, respectively referred to as American Viticultural Areas (AVA) and Geographical Indications (GI). Before the GI scheme, introduced as part of a wine agreement with the EU, Australia had established a Label Integrity Program, which is intended to ensure the correct designation of region and variety on the bottle label. The GI system has a hierarchy of States and Territories, Zones, Regions, and Subregions, based on soil, local climate, and history. The AVA system in California even goes so far as to call a region, such as the Napa Valley, an "appellation" and subregions within that, "subappellations."

But the identification of these geographical regions has contributed little to a scientific understanding of the relationship between particular wines and the "earthy component" of the *terroir* in which they are produced.

9.2.2   *The Technical Approach*

The technical approach avoids defining *terroir* in terms of traditional regions. Rather, this approach seeks to identify specific properties of the soil and environment that determine wine character. Of course, these properties will also influence a site's suitability to grow certain varieties, so to that extent *soil*, *environment*, and *variety* will interact to determine wine character.

The technical approach has not been very successful, primarily because it is difficult to unambiguously identify the causal relationships between soil properties and the complex chemistry that determines wine character. For example, Rankine et al. (1971, p. 33) in South Australia wrote, "there seems to be little, if any, detailed factual information on the influence of different soils on the composition of grapes . . . or the quality of wines made from the same varieties." These scientists examined the relationship among soil, grape juice composition, and wine quality in a series of experiments over six years in the Barossa and Eden Valleys, and on irrigated vineyards in the Riverland region of South Australia (box 9.2). Five soil types were included and a range of chemical analyses performed on the grape juices and wines. Unfortunately, no detailed soil analysis was done. The soils were differentiated by their general-purpose classification names, which convey little information about the soil's properties relevant to vine growth and grape com-

---

Box 9.2   *Comparison of Soil Types, Grapes, and Wine of South Australia*

The yield and quality of grapes were compared when grown on the following soil types

- Red Brown Earth ⎫
- Solonized Solonetz ⎬ Barossa Valley
- Yellow Podzolic Soil ⎫
- Grey Brown Podzolic Soil ⎬ Eden Valley
- Solonized Brown Soil   Riverland

These soils all had duplex profiles (section 1.3.2.1). The two Podzolic Soils were acidic, whereas the Red Brown Earth and Solonized Solonetz had a neutral topsoil over an alkaline subsoil, and the Solonized Brown Soil was alkaline throughout. The presence of Na^+ ions had influenced the profile development and structure of the two solonized soils (section 7.2.3). The varieties Riesling, Clare Riesling, and Shiraz (Syrah) were grown in the Barossa and Eden Valleys (rainfall 520–684 mm/year), and Clare Riesling and Shiraz were grown under irrigation in the Riverland (rainfall 528 mm, supplemented with 760–960 mm of irrigation, depending on the season). Seasonal rainfall was normal during the experiment (1958–1964), except for one very dry season.

---

position. Nevertheless, some effects of soil type on grape ripening and composition were observed, namely:

- Grapes ripened earlier on Red Brown Earths than on Solonized Solonetz in the Barossa Valley. This was attributed to the darker surface of the former soils, which absorbed more heat and created a better microenvironment for ripening.
- Riesling grapes grown on Yellow Podzolic soils in the Eden Valley (higher and cooler than the Barossa) had less titratable acidity and a higher juice pH than Riesling grapes grown on Grey Brown Podzolic soils in the same location.
- In addition to these compositional variations, yields on the Red Brown Earths and Solonized Solonetz of the Barossa Valley differed. These differences were attributed to variations in soil depth, water holding capacity, and drainage.

The composition, yield, and color intensity of grapes grown under irrigation in the Riverland were different from grapes of the same variety grown without irrigation in the Barossa and Eden Valleys, as shown in table 9.1. Wines made from the varieties at each site were evaluated by tasting panels. The "dryland" wines had a lower pH, lower K concentration, and more color (for Shiraz) than the "irrigated" wines. Dryland Riesling wines scored better for flavor and aroma. These results seem to confirm the long-held European view that wines made from low-yielding vines with a *regulated* water supply are superior to wines made from high-yielding vines liberally supplied with water. However, because the soils were not the same in the irrigated and nonirrigated areas, the influence of other soil properties on these differences cannot be dismissed.

Research in Bavaria (Wahl 2000) provides another example of the technical approach to defining the causal relationships between soil and wine character. A

Table 9.1   *Comparison of Yield, Grape Composition, and Wine Quality in Dryland and Irrigated Vineyards*

Dryland: Barossa Valley (Warm) and Eden Valley (Cool)	Irrigated: Berri and Loxton in the Riverland (Hot)
Yields (t/ha)	
2.5–7.5	15–25
Grape composition (Shiraz)	
Lower pH, lower K, lower Cl, higher titratable acidity (as tartaric acid)	Higher pH, higher K, higher Cl, lower titratable acidity (as tartaric acid)
Wine properties (Shiraz)	
Lower pH, higher titratable acidity (as tartaric acid) 50–400% more color than in Shiraz grapes grown under irrigation	Higher pH, lower titratable acidity (as tartaric acid) Less color than in Barossa and Eden Valley wines
Wine properties (Riesling)	
No consistent significant differences in pH or titratable acidity between regions, but better aroma and taste scores for the wines produced from dryland vines	

Source: Data from Rankine et al. (1971)

trial from 1977 to 1998 grew the varieties Muller-Thurgau and Silvaner in large, 1-m-deep containers filled with soil from seven different parent materials. To eliminate the climatic variable from *terroir*, the containers were placed in a typical vineyard in the Franken region of Germany, with the aim of isolating the soil effect on wine character. The wines from the trial were compared with wines made from the same varieties in the original locations whence the soils were collected. Although there were differences in grape yield, the effect of soil type on the sensorial traits of the wines was very limited.

One problem with this type of experiment is that the root environment in a confined cubic meter of soil would have been very different from that of roots free to explore a larger soil volume in the field. Further, the hydrologic regime of each soil would have been changed by its removal from the field and repacking in a container. This point is emphasized by the results of Campostrini et al. (1996) in the Vino Nobile de Montepulciano territory of Italy. Working on soils derived from calcareous parent materials, with variable drainage and erosion, they found little relationship between fruit quality and wine personality (or "typicity") and the *AWC* and air capacity of the soils. Obviously, there is more to determining the hydrologic influences of soil on vines than can be expressed in laboratory measurements of *AWC* and air capacity.

To date, the technical approach to defining *terroir* has not been particularly successful. However, the application of new techniques in remote and proximal sensing, which allow the rapid collection of data on soil, cultural methods, and yield at very high observation densities (section 8.2.1), offers the opportunity to better define quantitative relationships between the soil and wine quality. Meanwhile, the lack of success in quantifying the key factors of a *terroir* is an advantage for those vineyards and regions that have established reputations for the distinctive character of their wines, based on long experience. This may explain why wine quality in the EU is defined in terms of the region of the wine's origin (see box 1.1). By this definition, *terroir* and wine quality become inextricably associated.

9.2.3   *An Alternative to Terroir*

There are several interpretations of wine quality and whether quality connotes *terroir* (Penn 2001). Many in the New World believe wine quality is determined by selected fruit properties and wine-making skills, and they largely ignore the influence of *terroir* (soil and climate) on wine character. Instead, the focus is on measuring those grape properties that appear to correlate directly with wine sensory properties. The concentration of anthocyanin pigments and total glycosol-glucose compounds in grapes is believed to indicate the intensity of wine flavors (Gishen et al. 2001). To be used as a quality criterion, these properties must be measured rapidly at the point of delivery to the winery. To this end, in Australia's Riverland and Sunraysia regions, BRL Hardy has used anthocyanin concentrations in red grapes, measured by Near Infrared Spectroscopy (NIRS), to assess grape quality and to determine payments to growers (Kennedy 2001). When used in this way, NIRS is essentially a "black box" method that must be calibrated against laboratory measurements of anthocyanin concentrations, as well as pH and °Brix, using very large sample sizes. The major problem is how to relate the NIRS results back to vineyard management, so that growers know which practices will produce the quality of grapes sought by the winery.

9.2.4   *Examples of the Expression of Terroir*

The objective approach to wine quality discussed in section 9.2.3 is a tacit admission that generic relationships between soil and wine character are unlikely to apply widely across regions, let alone across countries in different hemispheres. For this reason, to investigate more closely whether the soil factor of individual *terroirs* is important in determining the distinctive character of a wine, we must examine the evidence from specific regions in different parts of the world.

9.3   *The Bordeaux Region*

9.3.1   *Location, Climate, and Geology*

There are more than 100,000 ha of vines in the Bordelais, the region surrounding the city of Bordeaux and located on either side of the Gironde estuary along the lower reaches of the Garonne and Dordogne Rivers (fig. 9.1). To the east of the Dordogne is the St. Emilion-Pomerol region, between the rivers is Entre-deux-Mers, and to the west of the Garonne lie the districts of Médoc, Graves, and Sauternes. The land is relatively flat, ranging in elevation from sea level to about 120 m, and the climate is tempered, especially in the Médoc, by proximity to the sea. The mean annual rainfall in the city of Bordeaux is 920 mm, but the range from year to year can be large (e.g., 1160 mm in 1960 to 450 mm in 1989). Because the AOC vineyards are not irrigated, this variability in rainfall means that,

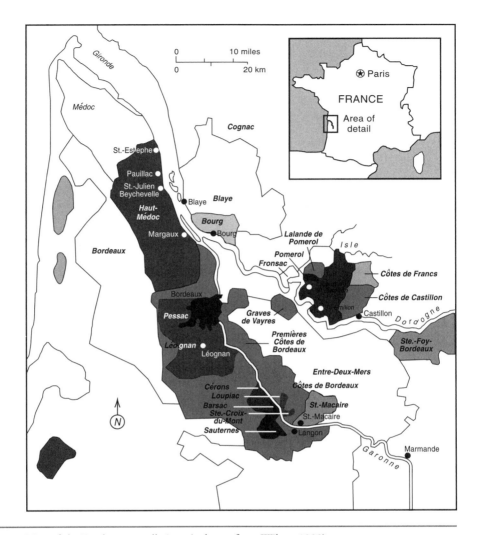

Figure 9.1   Map of the Bordeaux appellations (redrawn from Wilson 1998).

in dry years, the water supply from the soil is very important. Good drainage is also important in normal to wet years.

The rocks and sedimentary deposits of the region are of Tertiary and Quaternary age. There were several marine deposition phases during the Eocene through Miocene epochs (see table B1.2.1). The most important of these deposits in the Bordeaux region was the "starfish limestone"—Calcaire à Astéries—of Oligocene age, which outcrops in places such as in the center of the St. Emilion appellation (fig. 9.2). But this is primarily a region of molassic and alluvial deposits of the late Pliocene through to glacial and interglacial periods of the Pleistocene. There are extensive fans and terraces of alluvial gravels and sands with variable clay content, whose physical composition, elevation, and thickness largely determine the quality of the soils for viticulture.

Figure 9.2   Grand and Premier Cru Classé vines on the slopes of the molasse du Fronsadais, under the escarpment of the Calcaire à Astéries plateau, St. Emilion. Photograph by the author. See color insert.

9.3.2   *The Soils and Wines*

Given the nature of the parent materials, the soil of the Bordeaux region is extremely variable. References to the soils of St. Emilion-Pomerol, on the one hand, and those of the Médoc, on the other, give an indication of how soil properties influence wine character and quality in the region.

St. Emilion-Pomerol. The vines, predominantly Merlot, Cabinet Franc, and Cabernet Sauvignon, are normally planted at 6000 per ha on soils that are slightly acid to alkaline (up to pH 8). The soils on the low plateaus are Rendzinas and shallow Calcareous Brown Soils (<1 m deep). Immediately below the limestone escarpment, deeper Calcareous Brown Soils have formed on slopes where calcareous, marly deposits outcrop (molasse du Fronsadais) (van Leeuwen 1989). The surface of the plateau soils has also been enriched to some degree with silt-size material or *loess* (section 1.3.1.2), blown by the wind in the late glacial phases of the Pleistocene. The loess has replaced some of minerals leached from the older soil profiles and may partly explain why the best grapes in St. Emilion are grown on these soils (Enjalbert 1983). The textures are generally loams to clay loams. Although the soils on limestone are shallow, the vine roots penetrate deeply via weathering fissures in the rock and are also supplied with water by capillarity. The water-release characteristic of the limestone is such that there is good drainage at high matric potentials (ψ_m 0 to -10 kPa) and hence a good air capacity. But there is also a substantial amount of readily available water to be released between -10 and -60 kPa (section 6.5.1). The water table is generally very deep in the limestone.

Of the 13 First Growth wines in St. Emilion, 8 are produced from vines on the plateaus and molassic slopes. The yields are low and the grapes ripen slowly, but the wines do not have aggressive tannins and age very well.

The soils on the lower land of St. Emilion, and most of Pomerol, fall into three broad categories (van Leeuwen and Merouge 1998):

1. Soils that are gravelly (50–70%) throughout the profile and very well drained
2. Soils that are sandy in the topsoil (up to 1 m deep), but with increasing silt and clay content (ca. 20%) in the subsoil
3. Soils that have a sandy to sandy clay texture in the topsoil (30–50 cm deep), but become very high in silt and clay (>90%) in the subsoil. Often this subsoil layer is compacted.

There are many variations within these categories, particularly with respect to the proportions of gravel and coarse sand in the topsoil and the interleaving of sand and clay "lenses" in the subsoil. Examples of these main profile types are given in figure 9.3a–c. Soils of type 2 frequently have a permanent water table within 1 to 2.5 m of the surface. Surprisingly, soils of type 3 do not develop perched water tables over the clay subsoil. Detailed studies by French scientists at the University of Bordeaux and INRA have shown that the quality of the grapes and wine produced is primarily determined by the water supply to the vines (Seguin 1986). On the gravelly and clayey soils (types 1 and 3), the vines showed a gradual increase in water stress from veraison on, which was associated with a higher fruit sugar concentration, lower malic acid levels, and more concentrated anthocyanins and total polyphenols (tannins) at vintage. However, on the sandy soils with a permanent water table, which remained within reach of the roots during the entire growing season, the vines did not show any water stress. Berry weights were high and favorable indices, such as anthocyanin and tannin concentrations, were low. The results of these measurements are summarized in table 9.2. The wine made from the grapes was ranked by tasters in the order type 3 > type 1 > type 2, and a number of vineyards on type 3 soils produce wines that are Grands Crus Classés.

The overriding influence of soil type on the personality and quality of wines in St. Emilion-Pomerol is through the regulation of the water supply to the vines, especially in the post-veraison period. For this reason, drainage with pipes at 80-cm depth and 10-m spacing is used to remove excess water from clay subsoils, and groundwater pumping is used to control the water table in the deep sandy clays.

The Médoc. The Médoc is a long strip of land protected in the west from Bay of Biscay winds by extensive pine forests, and adjoining the estuary of the Gironde in the east. The soils are formed on Quaternary alluvial and glaciofluvial deposits that are deep and gravelly in the south (Haut-Médoc) and increase in clay content to the north (Central and Bas-Médoc). The vines are planted at 10,000 per ha, with the best vineyards situated on well-drained soils and facing east or northeast (where they "see the river"). A number of châteaus are classed as Premier Cru (First Growth) in Margaux in Haut-Médoc and as First and Second Growths in the communes of St. Estephe, Pauillac, and St. Julien in the Central Médoc.

Because the soils were formed on alluvial deposits, the soil texture is highly variable over short distances, both laterally and vertically, and drainage is very dependent on local relief. Records from the seventeenth and eighteenth centuries of Château Latour in Pauillac reveal that the land was originally divided into parcels of 1–5 ha according to soil type, particularly the proportion of gravel in the pro-

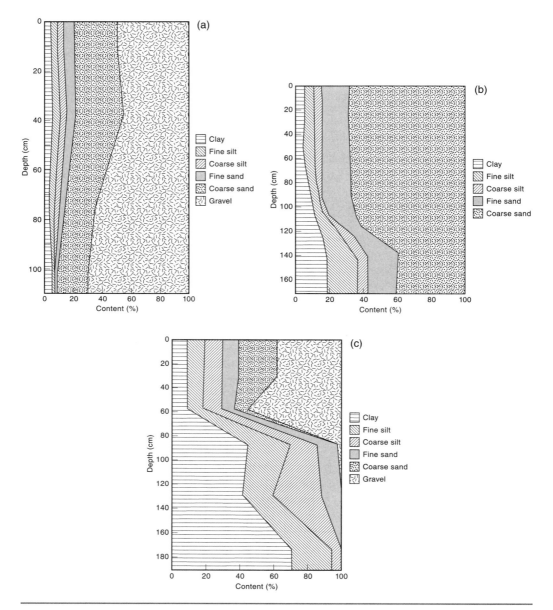

Figure 9.3   Gravel and fine earth contents of soil profiles in the lower land of St. Emilion-Pomerol. (a) Deep gravelly soil (type 1) at La Medoquine, Château Cheval Blanc. (b) Deep sand over sandy clay (type 2) at Haut Dominique, Château Cheval Blanc. (c) Gravelly sand over heavy clay (type 3) at Château Trottany in Pomerol (redrawn from van Leeuwen and Merouge 1998).

file, the size of pebbles on the surface, drainage, and aspect (Johnson 1994). As in St. Emilion-Pomerol, the depth of rooting and regulation of the vine's water supply are the main determinants of quality and wine character. Vines on shallow, very gravelly soils, as in Margaux, may suffer too much water stress during dry summers but do very well in wet years when the free drainage of the gravelly

Table 9.2   ***Grape and Wine Quality Factors Associated with Three Common Soil Types
in St. Emilion-Pomerol***

Grape or Wine Property	Deep Gravelly Soil, Very Well Drained Soil Type 1	Deep Sandy Topsoil Over Sandy Clay Subsoil, with a Water Table Soil Type 2	Shallow Sandy Topsoil Over Very Heavy Clay Subsoil, Gley Features Soil Type 3
Total acidity (mmol charge/L)	58	62	64
pH	3.87	3.74	3.80
Alcohol concentration (%)	12.3	12.4	12.8
Anthocyanins (mg/L)	510	350	510
Total phenols (D280)[a]	41	35	49
Total tannins (g/L)	2.0	1.7	2.4

[a]Measured by absorbance of light of 280-nm wavelength
Source: Data from van Leeuwen et al. (1998)

soils is an asset. In Central Médoc, where most of the First and Second Growths
are produced, the soils are deeper and contain more silt and clay. The best wines
are produced from the tops and upper slopes of the gravelly mounds (up to 30 m
high in Pauillac), whereas the second and lower order growths are produced from
the lower slopes and hollows where the soil does not drain as well. Not only does
the depth to the permanent water table vary with the topography (and season),
but perched water tables can develop over lenses of impermeable clay at different
depths in the profile, as shown in figure 9.4.

Figure 9.4   Water movement shown in the cross section of a gravel mound in the Médoc (redrawn
from Wilson 1998).

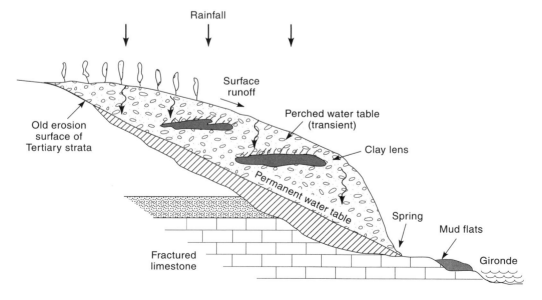

9.4   *The Burgundy Region*

9.4.1   *Location, Climate, and Geology*

The vineyards of Burgundy (Bourgogne) extend along the eastern escarpment and footslopes of the Massif Central from Dijon in the north to Lyon in the south. The best slopes face the east, southeast, and south, overlooking the valley of the Saône. The climate is continental, with an average rainfall of 700 mm and a range from 450–950 mm within the region. The three main subregions are the Côte d'Or, Mâconnais, and Beaujolais (fig. 9.5). North of Mâcon, the underlying rocks are hard limestones, interbedded with softer limestones and marl (a mixed, fragmented limestone-clay deposit) of Jurassic age. South of Mâcon, in Beaujolais, the rocks are mainly granite, which, as it was intruded through the overlying volcanic lavas and ash, metamorphosed these rocks to form schists (Wilson 1998). The escarpment, which is between 225 and 350 m high, is formed by a major fault line that trends mainly north–south. This fault is crossed in several places by minor

Figure 9.5   Map of the geology of the Burgundy region (redrawn from Wilson 1998).

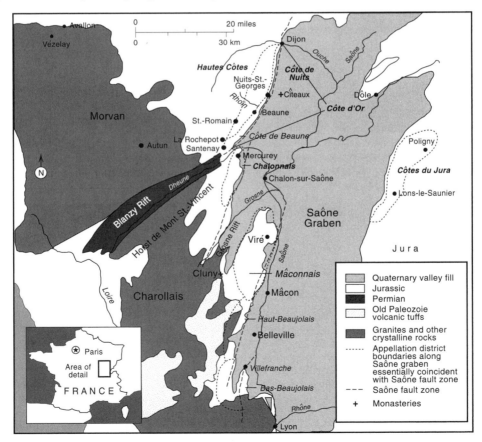

east–west faults that have been the nucleus for the formation of small valleys (combes) through weathering and erosion.

Wine making in Burgundy dates from at least the twelfth century when Cistercian and Bendictine monks applied the concept of *terroir* in small vineyards of the Côte d'Or. During subsequent centuries, the accumulated empirical knowledge of the relationship between individual *terroirs* and the quality and character of their wines has led to the recognition of Grand Crus, Premier Crus, and Village categories of wine. Planting density in the Côte d'Or is usually 10,000 vines per ha. The vigneron aims to grow small bunches and berries, with a limited number of shoots and bunches per meter of shoot, to produce a balanced vine (see box 9.1). For AOC wines, irrigation is not permitted, and yields are restricted to <5 t/ha. As in the Bordeaux region, the key soil properties that determine wine quality are slope, aspect, stoniness, and clay content, which influence primarily drainage during winter and spring, and water retention and release during summer.

9.4.2   ***The Soils and Wines***

Côte d'Or. This subregion is the home of the Pinot Noir grape, which is grown predominantly in the communes of the Côte de Nuits on relatively shallow soils, the Rendzinas and Calcareous Brown Soils formed on hard crinoidal limestone (Middle Bajocian) and the "oyster marl" (Wilson 1998). The pH is neutral to alkaline. Under native forest, these soils were rich in organic matter, as seen in figure 9.6. But after long periods of cultivation, much of the organic-rich A horizon

Figure 9.6   A Calcic Humifère Brunisol (Humic Calcareous Brown Soil) under forest in the Côte d'Or. Photograph by the author. See color insert.

Figure 9.7   Fractured rock of Middle Jurassic limestone underlying Grand Cru vineyards near Gevrey-Chambertin, Côte d'Or (scale 15 cm). Photograph by the author. See color insert.

has eroded away. Over the years, the vignerons have carried eroded sediment from the lower slopes back to the upper slopes.

The most favored areas are at the midslope, where the gradient is 3–5%, rising in places to 20%, and the soils are stony (5–40% gravel-size fragments). Although the Rendzinas and Calcareous Brown Soils on limestone are generally <1 m deep, the vine roots can penetrate more deeply through fissures in the fractured rock below, as shown in figure 9.7. The limestone provides a regulated supply of water when the vines come under stress in late summer, in a way similar to that described for the vines on limestone in St. Emilion (section 9.3.2). Wines of lesser quality are produced on the upper escarpment—the Hautes Côtes—and on the lower slopes where the soils formed on colluvial limestone and marl are deeper, less well drained, and higher in nutrients.

Both red and white grape varieties are grown in the southern part of the Côte d'Or—the Côte de Beaune—but the most favored variety is Chardonnay. It grows best on the crumbly marl and limestone-derived soils, especially in the communes of Puligny-Montrachet and Chassagne-Montrachet.

Mâconnais. This subregion, of lesser distinction than the Côte d'Or to the north and Beaujolais to the south, has two main terrains:

- Sandstones and other siliceous deposits, supporting Leached Brown Soils of pH 5–6. The main grape variety grown on these soils is Gamay, which generally produces wines of "grand ordinaire" quality.

- Limestone outcrops supporting Rendzinas and Calcareous Brown Soils (pH 7 and above), on which Chardonnay is grown. This terrain produces especially distinctive wines in the Pouilly-Fuissé area.

Beaujolais. This is an extensive area of vineyards south of Mâcon, which is divided into well-drained acidic Brown Soils on granite hills north of Villefranche-sur-Saône (Haut-Beaujolais) and poorly drained clay soils to the south (Bas-Beaujolais). The grape variety is almost universally Gamay, and the region is famous for its Beaujolais Nouveau, released for immediate consumption on the third Thursday of every November.

The best wines (Cru Beaujolais) are produced on the steep hills (up to 600 m high) of the north, between Côte de Brouilly and St. Armour Bellevue, on shallow coarse-textured soils over granite and schist. Here, the Gamay grape, a potentially vigorous variety, is restrained in its growth and produces wines of subtle flavors. Variation in the metal content (e.g., manganese) of the granite and schist is said to account for the individuality of the wines in this area (Wilson 1998).

9.5 *The Napa Valley of California*

9.5.1 *Location, Climate, and Geology*

The Napa Valley Viticultural Area, outlined in figure 9.8, occupies some 120,000 ha, of which ca. 16,000 ha are planted to vines. The area is situated just to the north of San Pablo Bay near San Francisco. By world standards, Napa is small, but it has a reputation for producing quality wines of distinctive character. Much of this distinction is attributed to the range of mesoclimates, which become warmer from south to north, and the soil types. California is the place where the Heat Summation system of classifying climates for viticulture was developed (appendix 14). The rainfall ranges from ca. 750 to 1500 mm, depending on position in the valley and elevation, and it falls mostly between November and April. The main determinants of soil variation are geology and relief.

The Napa Valley is relatively young, having been formed by faulting and folding of old marine sediments (>65 M yr B.P.) in Miocene times, followed by volcanic eruptions, which produced basaltic to rhyolytic lava flows and tephra, over the last 5 M yr (Elliot-Fisk 1993). The resultant structure is a giant syncline in which ultrabasic rocks of the ancient oceanic crust are overlain by more recent marine sediments, and these by the younger volcanic deposits and their erosion products. As shown by the east–west section through the Napa Valley in figure 9.9, the oldest rocks outcrop high in the ranges on either side (up to 1350 m), and the younger rocks outcrop progressively at lower elevations. Alluvial and colluvial fans (locally called "benches") occupy parts of the lower footslopes, and deep alluvial clays, silts, sand, and gravel deposited by the Napa River cover the valley floor.

9.5.2 *The Soils and Wines*

Vineyards in the Napa are mainly confined to the fertile alluvial soil of the valley floor and the fans or benches extending up the slopes. Vines do not grow well on

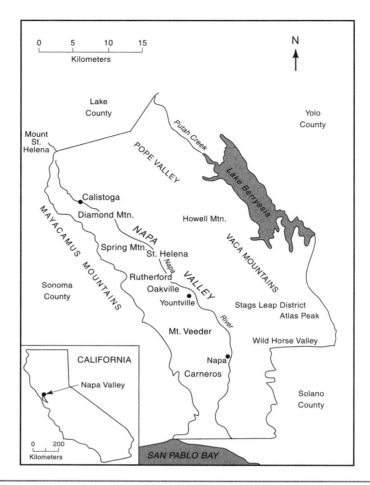

Figure 9.8   Map of the Napa Valley Viticultural Area with its constituent AVAs or subappellations. Both the AVA name and Napa Valley name must appear on bottles of wine in which 85% or more of the grapes come from that AVA (Elliot-Fisk 1993). Reproduced from the Journal of Wine Research with permission of Taylor & Francis Ltd. The journal's Web site is located at http://www.tandf.co.uk/journals.

the soil formed on the ultrabasic rocks high in Mg and sometimes Ni, which is toxic. Increasingly, however, vines are being established on soil formed on the volcanic ash or tuff. Vines grow well on the alluvial valley floor, but they can be excessively vigorous. Vigor is managed by the appropriate choice of rootstocks (e.g., 101-14, 3309, and True SO-4), canopy pruning, and *RDI* (section 6.5.2). In the southern Carneros district, because the silty and clayey soils are naturally poorly drained, subsoil drainage has been installed to improve vine performance. Because of cooling winds from San Pablo Bay, Pinot Noir and Chardonnay grow well.

The most favored areas of Napa are the benches in the footslope regions, where the soils are shallower, less fertile, very stony in places, and well drained (Lambert and Kashiwagi 1978). Here, the mesoclimate is also very favorable, especially on the western side that receives the morning sun, but is shaded by the

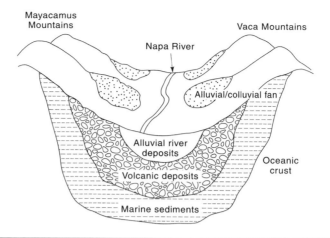

Figure 9.9   An east-west section through the strata of the Napa Valley, California (redrawn from Livingston 1998).

Mayacamas Mountains in the afternoon. Yields are low (ca. 5 t/ha), and the degree of stress suffered by the vines as the fruit matures produces intense flavors. On the western side, the Rutherford AVA is renowned for its Cabernet Sauvignon, and distinctive wines are also produced in the Oakville and St. Helena AVAs to the south and north, respectively. Similarly, Cabernet Sauvignon wines from the Stag's Leap AVA on the eastern side have a distinctive character attributable to the volcanic parent material of the alluvial and colluvial soils found there (Livingston 1998).

The emphasis on soil as an important component of the *terroir* of ultra-premium and superpremium Napa Valley wines has come fairly recently. Previously, Winkler et al. (1974, p. 74) wrote that "varieties of the highest quality produce excellent wines when grown on a number of different soil types. Thus, the difference in the character of wines can hardly be attributed to specific soil types." These authors argued that the principal factor controlling fruit quality in California was the mesoclimate (expressed by heat summation), through its influence on the sugar:acid ratio, total acidity, tannin content, and other minor constituents. Grape variety was the determinant of inherent characteristics, such as aroma and flavor. But the current view is that quality depends on how well the grapes are grown and the wine made, whereas the character of a particular varietal wine that has been well made depends on soil type and other factors such as mesoclimate and topography, including aspect (Elliot-Fisk 1993).

9.6   *The Willamette Valley of Oregon*

9.6.1   *Location, Climate, and Geology*

The Willamette Valley AVA is relatively small, production-wise, but it has established a reputation for its Pinot Noir wines. Chardonnay, Riesling, and Pinot Gris are also grown. In fact, partly because of its focus on Pinot Noir and climatic similarities, this area has sometimes been called the "Burgundy of the West" (Darlington 1999). Since the valley is sandwiched between the Coast Ranges to the west and the Cascades to the east, the summers are warm, but the winters are cold and wet (rainfall ranging from 1000 mm on the valley floor and in the south to 1500 mm in the north and on higher areas). A strong French influence is evident in the choice of vine clones, planting density, trellising, and pruning.

9.6.2   *The Soils and Wines*

Unlike in Burgundy, soil variation in the Willamette Valley is comparatively simple. There are two main soil types:

- The *Jory soil* has formed on basaltic colluvium, and typically occurs on the lower rolling foothills with slopes of 2–60%. It is a very deep (up to 2.5 m) and well-drained soil, of silty clay loam to silty clay texture in the A horizon, grading into a deep, well-structured B horizon of reddish brown clay. The profile is acid throughout, ranging from pH 5.6 (in water) at the surface to pH 5.3 at depth. Shallower soils of the Nekia and Bellpine series occur on the rounded foothills and abrupt steep breaks, often on colluvium from basalt, tuff, or sedimentary rocks.
- The *Willakenzie soil* is a moderately deep (up to 90 cm) soil, formed on colluvium and residual weathered sandstone or siltstone. The texture is silty clay loam throughout the profile, but there can be an impediment to vertical water flow at the top of the B horizon and also at the parent material interface. The pH changes from 6 in the A horizon to 4.7 in the B horizon.

Generally, the soil water supply is adequate on the Jory, although in some cases too much water in spring and early summer promotes excess vigor. Many vineyards on the Jory are without irrigation, and cover crops are used to reduce vigor and soil erosion. The growers plant in close spacing (1.5 × 1 m) and pull fruit to control yields. Water management is more of a problem on the Willakenzie, Nekia, and Bellpine soils. Perched water tables may develop in winter in the Willakenzie, but there is often a lack of soil water in dry summers, because of poor root penetration into the subsoil. Irrigation is used to advantage here. A number of growers are conscious of environmental concerns about wine production and lean toward minimum inputs (i.e., low nutrients and minimal use of pesticides), while not wanting to be wholly "organic." Thus, they sow a legume in the cover crop, make and apply compost, and spray soft rock phosphates onto the vines. Lime is applied to raise the soil pH to 5.5, as necessary.

The relatively uncomplicated soil distribution and the high level of nutrient management practiced by the growers may be reasons why the soil is not often identified as a major factor in the *terroir* of the Willamette Valley. Nevertheless, as with the granites and vulcanized schists of Haut-Beaujolais and the Cambrian volcanics of Heathcote, Victoria (section 9.7.3), some believe that particular minerals derived from the basalt confer a distinctive character to the grapes grown on Jory soil (Darlington 1999).

Selected Wine Regions of Southeastern Australia

As indicated in section 1.1, soil is generally not considered to be an important component of *terroir* in Australia. Summarizing Australian opinion, Halliday (1993) considered that no link had been established between soil mineral composition and grape quality or wine character, with two exceptions: a certain amount of N is essential, and excess K can be a problem, especially in warmer regions. In this, Halliday was also echoing the conclusions of Seguin (1986, p. 871) who wrote, "in the Bordeaux area the chemical properties of soils . . . do not have a definite influence on the quality of harvests and wines. The quality of terroirs is perhaps better explained by considering the physical properties . . . and their consequences for root development and on the regulation of water supply to the vine."

With this background, it is no surprise that there is little knowledge of the specific role of soil in the *terroir* of Australian wine regions. However, for the purposes of marketing and identifying an exclusive GI, there are regions that now claim a distinctive influence of the local soil on the character of the wines. Three of the more important such regions are briefly discussed next.

9.7.1   ***The Coonawarra Region of South Australia***

The Coonawarra Region is flat, with an elevation of 55–60 m above sea level. The annual rainfall is ca. 650 mm, with a marked winter incidence. Summers are hot and dry, and there is a significant temperature gradient from south to north (grapes ripen two weeks earlier in the north). Nevertheless, temperature extremes are moderated by proximity to the cool Southern Ocean. Here, soil and climate have combined to create a *terroir* that confers a distinctive character on quality wines. Today, Coonawarra makes much of the influence of its "unique" Terra Rossa soil on its superpremium and premium red wines (51% Cabernet Sauvignon and 21% Shiraz).

The Terra Rossa soil has formed on a sandy, porous limestone laid down during the Miocene epoch (22 M yr B.P.). During the Pleistocene, fluctuations in sea level associated with the glacial and interglacial periods led to a succession of sandy beaches being formed over limestone ridges. These have subsequently been exposed as the land was uplifted, and the sea has retreated to the west. The lower-lying areas between the ridges are poorly drained and swampy, in contrast to the higher ridges, which are well drained. Although the boundaries of the much-prized Coonawarra GI are still in dispute, the region is currently a cigar-shaped strip of land, about 20 km long and 5 km across at its widest point, between the towns of Coonawarra and Penola in the southeast of South Australia. Grapes are grown on other outcrops of limestone with similar soil in the wider region, so that the whole GI Zone is called the Limestone Coast.

A catena of soils (section 1.3.4.1) has formed on the limestone ridges and old beaches. During the last glacial period, silt-size particles of clay, quartz, and limestone were blown from the dry lake beds and shoreline to accumulate on the ridges. With weathering of the limestone by dissolution in Recent time, this loessal material has accumulated and given rise to the sequence of soils shown in figure 9.10. Lowest in the sequence are very poorly drained Black Earths, which are used only for pasture. Next come the Groundwater Rendzinas on which vineyards

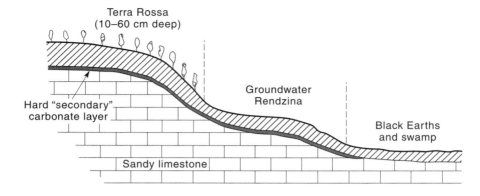

Figure 9.10   A catenary soil sequence in the Coonawarra Region on Miocene limestone, influenced by Pleistocene processes (courtesy of R. W. Fitzpatrick).

have been established during recent years, but which may have drainage problems after a run of wet years. The best soil is the Terra Rossa on the high ridge. Its profile is similar to a Brown Calcareous soil, that is, deeper than that of a Rendzina but with a pronounced red color due to residual ferric oxide impurities from the weathering limestone and loess (fig. 9.11). The soil is well structured, with a clay loam texture that provides an *AWC* of 17 mm per 0.1 m depth. However, vine roots can penetrate the porous and fractured limestone below and extract water at a relatively slow rate, in a way similar to that of vines grown on the lime-

Figure 9.11   A Terra Rossa soil profile in the Coonawarra Region, South Australia. Note the deep solution hole in the limestone, filled in with red soil. Photograph by the author. See color insert.

stone soils of St. Emilion and the Côte d'Or. Thus, good drainage during the winter and regulation of the water supply to vines during the summer are the salient soil factors governing fruit quality and wine character in the Coonawarra.

9.7.2   *The Barossa Valley of South Australia*

The Barossa Valley, probably the best known wine region in Australia, has a wine-making tradition going back to the early 1840s. With the contiguous Eden Valley, it comprises the GI Barossa Zone. The Barossa Valley occupies the land on the western slopes of the Barossa Range and the valley below, to the northeast of Adelaide. The average elevation is 274 m, and the annual rainfall is between 500 and 550 mm. The Eden Valley lies to the east, at an altitude >400 m. The Zone grows red grapes, mainly of the varieties Shiraz (Syrah), Grenache, and Cabernet Sauvignon, and white grapes of the varieties Semillon, Riesling, and Chardonnay. The Eden Valley produces wines more in the cool climate style, including Shiraz, Riesling, Chardonnay, and Pinot Noir.

The rocks of the Barossa Range consist of a belt of metamorphosed Cambrian limestone within the gneisses and granites that form the higher elevations. The valley is filled with colluvial and alluvial sediments derived from these rocks. By virtue of their mode of deposition and the variety of surrounding rock types, the Barossa Valley soils are extremely variable (compare to the Napa Valley). According to Northcote et al. (1954–59), the soils at the lowest elevations are mainly grey and brown clays that are not well suited to viticulture. At slightly higher elevations, typically in undulating country, the dominant soils are deep Red Brown Earths (with calcareous subsoils) interspersed with Solonized Solonetz and other duplex soils with sodic features (see box 9.2). The Red Brown Earths are the most favored soils for viticulture, with vines on the Solonetz soils suffering water stress unless they are irrigated.

Partly because of the complex soil pattern and the practice of large corporate wineries producing crossregional blends to ensure consistent varietal wines, the soil has not been featured as an important component of the Barossa Valley terroir. However, the Barossa Valley is renowned for its robust red wines, in particular those made from Shiraz grapes, such as Penfolds Grange (formerly Grange Hermitage). The key factors in the quality of fruit and rich, distinctive flavors of the wines, particularly those made from Shiraz and Grenache, appear to be the age of the vines (up to 100 years), and the absence of irrigation (see table 9.1). When the vines rely solely on the supply of soil water, vigor is restrained, yields are relatively low (>7.5 t/ha), and fruit quality high.

9.7.3   *The Heathcote and Nagambie Regions of Victoria*

The Heathcote Region is a small area in central Victoria that has some similarities with the Barossa Valley, because it has a reputation for intensely flavored Shiraz and Riesling wines. The annual rainfall is 550–575 mm, with cold wet winters and hot dry summers. However, the distinctive feature of many of the vineyards is the deep, well-structured Red Cambrian Soil (fig. 9.12). The Cambrian rocks (ca. 550 million years old) are a volcanic intrusion, only a few kilometers wide, which consists of basic dolerite and diabase outcropping with cherts and conglomerates. The most highly prized wines are produced from unirrigated vines, such as at Jasper Hill, where the target yield is only 1.2–2.5 t/ha.

Figure 9.12   Profile of a Red Cambrian Soil (Krasnozem) under vines at Heathcote, central Victoria (scale 15 cm). Photograph by the author. See color insert.

Approximately 50 km to the east of Heathcote is the Nagambie Lakes Sub-region. The soils here are mainly alluvial sandy loams over clays, deposited in a complex pattern by the Goulburn River. Chateau Tahbilk Estate makes some wine from Shiraz vines that are 141 years old. These vines are growing on a well-drained sandy ridge (fig. 9.13), which enabled them to survive the phylloxera invasion that occurred in the late nineteenth century. As in many other areas, the best wines tend to come from the oldest vines—typically those that are deep-rooted and which naturally achieve good balance year after year.

9.8   *Concluding Remarks*

The overriding conclusion to be drawn from the preceding chapters is that vineyard soils around the world are exceedingly diverse. Even where there is a common parent material, such as the much-favored limestone, the soils are variable, depending on the following factors:

- The permeability and degree of fracturing of the rock, which influences the rate of water percolation and ease of root penetration
- The quantity and nature of non–calcium carbonate impurities in the limestone (various forms of SiO_2, Fe and Al oxides, and clay minerals)
- The prevailing climate and site topography, which influence the rate of weathering of the limestone and leaching of weathering products

Figure 9.13   Very old Shiraz vines on a sandy ridge at Chateau Tahbilk, Nagamie Lakes Sub-region, central Victoria. Photograph by the author. See color insert.

• The vegetation, strongly determined by climate, but which also influences soil formation through root action and the return of litter
• The length of time the limestone has been weathering, which influences the depth of accumulation of resistant mineral residues and humified organic matter

The important effects of climate (macro-, meso-, and microclimate) on vine growth and fruit ripening have been convincingly argued by Gladstones (1992). The rates of growth and ripening strongly influence fruit quality and wine character through the ratio of sugars to acidity, juice pH, the accumulation of nonvolatile flavonoids (condensed phenolic compounds), and the more volatile compounds responsible for subtle fruit flavors and aromas. Others such as Smart and Robinson (1991) assert that control of the microclimate by canopy management is especially important in regulating the vine's physiological processes to achieve good fruit quality and wine character.

However, we can also see from the evidence of Burgundy, Bordeaux, the Napa Valley, and Coonawarra, that the soil is an important component of a vineyard's *terroir*. To identify the key relationship between the soil and wine character in particular, few would go to the extreme of "tasting the soil"—said to have been done by the Cistercian monks of the Clos de Vougeot in the Côte d'Or (Johnson 1994). The complex relationships between a soil's physiochemical and biological processes and the chemistry of the berries, fermentation of the must, and maturation of the wine is unlikely to be elucidated in the immediate future. But certain general conclusions about the influence of soil on wine character and quality can be drawn, as follows:

- *The thermal regime of the vine's microclimate is influenced by the soil,* through surface color, stoniness of the surface (affecting heat absorption and conduction during the day, and reradiation at night), and clay content (affecting water holding capacity, which in turn affects the soil's heat capacity). Heat storage by the soil is particularly important in cool-climate vineyards where the maintenance of warm soil into the autumn is necessary for root growth and the continued synthesis of cytokinins and their supply to the ripening fruit. In the ripening berries, cytokinins are involved in the accumulation of sugars, which are transported from the leaves. Sugars in excess of the amount needed for respiration and cell growth are used in the synthesis of flavonoid compounds that are associated with skin color, flavor, and aromas.

- *The soil water supply is important.* The key factors are the volume of soil exploited by the roots (mainly determined by rooting depth and planting density), the rate of water supply to the roots at ψ_m values between -10 and -400 kPa, and the rate of drainage of surplus water from the root zone ($\psi_m > -10$ kPa). In dryland vineyards, the soil's structure and texture and the depth to groundwater control the water supply. Optimal conditions have been identified, such as a soil strength <2 MPa at the field capacity (section 7.1.2), an infiltration rate >50 mm/hr (section 7.1.1), air capacity $>10\%$ (section 3.3.3), and a high value of *RAW* plus *DAW* (120–130 mm/m depth) (table 6.3). However, the influence of these soil conditions on the vine is modified by the underlying parent material in the case of older vines growing on porous and fractured rock, such as limestone in the Côte d'Or, St. Emilion, or the Coonawarra, or on schists in Beaujolais and the Hérault district of Languedoc-Roussillon. Furthermore, the soil water supply can be augmented by irrigation and manipulated by *RDI* or *PRD* to minimize the problem of excess vigor. Experience in regions with hot summers in California and Australia suggests that the most critical period for regulating the water supply to enhance quality is after fruit set until veraison. But in the cooler regions of Europe, regulation of the water supply during the long ripening period after veraison is found to be more important.

- *Good drainage is important,* especially in spring and early summer. A wet soil at this time of year slows the rate of soil warming and inhibits root growth. The best vineyards are consistently found on freely drained soils, such as form on limestone, limestone marls, or chalk in the Côte d'Or, St. Emilion, Champagne, and the Coonawarra, or on alluvial gravels and sands, and colluvial deposits, as in St. Emilion-Pomerol, the Médoc, Napa Valley, Barossa Valley, Maipo Valley of Chile, and the Hawkes Bay and Marlborough regions of New Zealand. Soil drainage can be improved by pumping where water tables are high (e.g., the deep sandy clays in St. Emilion) or by underground pipes in heavy clay subsoils (e.g., Carneros, parts of Pomerol, and the Médoc).

- *A balanced nutrient supply is important,* in particular of N and K (the latter especially so in hot, dry climates). Too high a supply of mineral N during the period of most rapid uptake between flowering and veraison causes excessive canopy development. The consequences are potentially too many bunches, large berries, and overshading of bunches. Shaded leaves produce less sucrose, so that translocation of sugars to the ripening fruit is diminished, with consequent adverse effects on quality. Conversely, N deficiency in the fruit may lead to "stuck fermentations" and undesirable H_2S production during fermentation (section 5.4.1.4). The dynamics of K uptake by vines are similar to those of N, and K has been found to attain high concentrations in the fruit of canopies with excessive shading (i.e., vines out of balance). High K concentrations are associated with an increase in the ratio of malic to tartaric acids, which leads to an increase in pH when the young wine undergoes malo-lactic fermentation. However, K deficiency impairs photosynthesis (and sugar production) in the leaves, and a low K:N ratio may induce "false K deficiency" (section 5.4.3).

9.9   *Summary Points*

In this discussion on the relationships involving soil, *terroir* and wine character, the following main points were made:

- The concept of *terroir* evolved from the historical French tradition that wines of a consistent character and quality came from particular regions, down to specific vineyards and blocks within vineyards. This concept, widely accepted in Europe, has not had much influence in the New World.
- Where empirical historical evidence of a consistent association between a site and wine quality is lacking, or deemed irrelevant, some corporate wine makers are analyzing grapes at the point of delivery by rapid methods that are believed to correlate with wine sensory properties. The results are used to adjust remuneration to growers on the basis of quality.
- Because the term *wine quality* is used in several different ways, its meaning has become vague. With skilled wine-making techniques, it should be possible to make quality wine from disease-free, mature fruit of any desirable variety. But the distinctive character of this wine will depend on the *terroir* (soil and climate), provided this influence is not obscured by extraneous factors in the vineyard or the winery.
- The scientific and technical understanding of the influence of the soil in *terroir* is not as well advanced as that of climate, grape variety, and cultural methods. Generalizations are made about the importance of soil temperature, soil water supply, aspect, drainage, and nutrient balance to achieve "vine balance" and good ripening conditions. But because soil is so variable in the landscape, it is likely that quantitative relationships between soil properties and the yield and composition of grapes will only be elucidated on a local scale (a few hectares in area). This is consistent with the empirical evidence of the AOC system, especially as it applies to the Grand Crus of Burgundy and the First Growths of Bordeaux. It is also consistent with the concept of precision viticulture, as outlined in section 5.3.5.
- Examples exist of a significant influence of soil on wine character for particular grape varieties grown in St. Emilion-Pomerol, the Médoc, the Côte d'Or, Beaujolais, Napa Valley, and the Coonawarra Region. In these and other cases, deeprooted old vines that are not irrigated generally produce high quality wines of distinctive regional character.

Appendices

Appendix 1   *Great Soil Group Soils and World Soil Classes*
             Relevant to Viticultural Regions of the World

Classes at the highest level in a soil classification are the most inclusive. These classes are based on a central concept that broadly encompasses members of the class. Each defining concept may be based on a factor of soil formation, such as parent material or climate, on a pedogenic process, or on an easily observed soil property or set of properties.

The Great Soil Group Classification was developed from Russian studies of soil formation, which particularly emphasized the role of climate and vegetation on pedogenic processes. The classification underwent many revisions during the twentieth century, with one of the latest versions being published by Stace et al. (1968). Relatively little use has been made of the higher categories of Order and Suborder. The intermediate level of Great Soil Groups (which gives the classification its name) has proved most popular. A similar approach underlies the development of soil classes (soil units) in the Soil Map of the World (FAO-Unesco 1988). The soil units are mapped as associations at a scale of 1:5,000,000. The units have been grouped into 28 World Soil Classes on the basis of the inferred pedogenic processes such as gleying (section 1.3.3.2), salinization (7.2.2), and lessivage (section 1.3.2.1), which have relevance to soil use and management.

In table A1.1, the soils commonly used for viticulture in different parts of the world are identified within their Great Soils Groups and cross-correlated with a World Soil Class where possible. The underlying connotation or inferred pedogenic process is also briefly described.

251

Table A1.1   *Great Soil Groups, Their Central Concept, and Corresponding World Soil Classes Used for Vines*

Great Soil Group	Broad Central Concept	World Soil Class	Underlying Connotation
Lithosols Calcareous Sands Siliceous Sands Alluvial soils	No profile differentiation	Leptosols Fluvisols	Weakly developed shallow soils Alluvial deposits from moving water
Brown and Red Calcareous Soils Rendzinas	Weak profile development; some with dark surface colors due to organic matter	Calcisols Kastanzems Phaeozems	Accumulation of calcium carbonate Soils rich in organic matter and brown or chestnut in color Soils rich in organic matter with a dark color
Red Brown Earths Leached Brown Soils Solonized Solonetz	Mildly leached soils (not strongly acid), but with profile textural differentiation (duplex)	Cambisols Solonetz	Changed in color, structure, and consistence Strongly expressed effect of salt, especially Na^+ ions, on soil structure
Red Podzolic Soils Yellow Podzolic Soils Brown Podzolic Soils	Mildly to strongly acid, with strong texture contrast	Acrisols Luvisols Podzoluvisols	Strongly acid with low base status Clay accumulation through lessivage
Krasnozems Terra Rossas Red Earths Calcareous Red Earths	Predominantly with sesquioxides	Ferralsols Plinthosols Calcisols	Soils high in sesquioxides Mottled clay materials that harden on exposure

Source: After Stace et al. (1968) and FAO-Unesco (1988)

Appendix 2   *Measuring the Soil Microbial Biomass*

Methods for measuring microbial numbers or mass fall into two main groups:

1. Direct observations of organisms as they grow on agar plates. A general nutrient medium intended to promote the growth of a range of organisms, or one intended for a specific group, may be chosen. Staining with dyes is also used to identify specific groups of organisms.
2. Physiological or biochemical methods such as the extraction of adenosine triphosphate (ATP) or fumigation with chloroform ($CHCl_3$), followed by extraction.

Measurements in the second group tell nothing about the species of organisms, but give estimates of the biomass size. The most widely used method is to fumigate the sieved soil in $CHCl_3$ vapor (free of acetone) for 24 hours, flush to remove the vapor, and extract in 0.5M K_2SO_4 for 30 min. The $CHCl_3$ kills and ruptures the living organisms. The dissolved organic C (DOC) in the extract is measured and the quantity per kg soil is calculated. An unfumigated soil is also extracted to act as a control. Biomass C is then calculated from the equation

$$\text{Biomass C} = \frac{(\text{DOC in fumigated soil}) - (\text{DOC in unfumigated soil})}{K_{ec}}$$

where K_{ec} is an empirical factor measuring the efficiency of C extraction from the microbial population. K_{ec} values are between 0.3 and 0.5, depending on the soil type, but the generally accepted value is 0.35. Measurements are usually made on the top 15–25 cm of the soil profile, where the C substrates and organisms are concentrated.

Appendix 3   *pK Values and Buffering*

Acids are defined as potential proton (H^+ ion) donors, and bases as potential proton acceptors. In aqueous solutions, acids dissociate H^+ ions that are accepted by H_2O. For example, the dissociation of an acid HA in water is written

$$HA + H_2O \longleftrightarrow H_3O^+ + A^- \tag{A3.1}$$

The ion H_3O^+ is called the hydronium ion (normally written as H^+). The degree of dissociation of HA determines the strength of the acid, which is measured by the dissociation constant K_a, defined as

$$K_a = \frac{a_{H^+}\, a_{A^-}}{a_{HA}\, a_{H_2O}} \tag{A3.2}$$

where the terms on the right-hand side are *activities* (see box 4.4). The activity of water (a_{H_2O}) = 1, and in dilute solutions, the activities of the other species can be replaced by concentrations. Thus, equation A3.2 is normally written as

$$K = \frac{[H^+][A^-]}{[HA]} \tag{A3.3}$$

or converted to the negative log form (cf. pH), which is

$$pK = pH - \log_{10} \frac{[A^-]}{[HA]} \tag{A3.4}$$

The smaller the value of pK, the stronger the acid. Most soil acids (associated with organic matter and soil minerals) are weak acids.

If equation A3.4 is rearranged as

$$pH = pK - \log_{10} \frac{[HA]}{[A^-]} \tag{A3.5}$$

we see that when there are equal concentrations of undissociated acid HA and base A^-, the pH = pK (this is the half-neutralization point). At this point, the addition of small amounts of acid or base causes the least change in pH. This is therefore the point of maximum buffering of the solution. The *buffering capacity* of a solution is its resistance to pH change when acid or alkali is added. An example of pH buffering capacity is given in section 4.6.3.3.

Appendix 4   *Examples of How to Calculate* $N_{min.}$ *for Different Organic N Sources*

The instantaneous rate of net mineralization of soil organic N is given by

$$\frac{dN}{dt} = -kN \tag{A4.1}$$

The amount of mineral N released over a period of time is calculated using the integral form of equation A4.1, that is

$$N_t = N_o \exp(-kt) \tag{A4.2}$$

where N_o and N_t are the amounts of organic N in a defined volume of soil at time zero and time t, respectively. The amount of $N_{min.}$ produced over the time interval is therefore given by

$$N_{min.} = (N_o - N_t) \tag{A4.3}$$

$$= N_o(1 - \exp(-kt)) \tag{A4.4}$$

This equation should be used if $N_{min.}$ is to be calculated for a relatively short period, up to 3–4 months. An example follows. Suppose that for a green manure crop, the decay coefficient k is 1.0 yr^{-1} and 1.5 t dry matter/ha is incorporated (containing 40% C and a C:N ratio of 20). From this, we calculate that the value of N_o for the green manure is

$$(0.4 \times 1500/20) = 30 \text{ kg N/ha} \tag{A4.5}$$

From equation A4.4, we calculate that the value of $N_{min.}$ produced in 3 months is

$$N_{min.} = 0.22 \, N_o \tag{A4.6}$$

$$= 6.6 \text{ kg N/ha} \tag{A4.7}$$

In other words, in 3 months, about 22% of the green manure N should mineralize.

For longer time intervals, when a large amount of organic N is involved, $N_{min.}$ can be calculated using an approximate solution of equation A4.1, which does not involve an exponential function, that is,

$$N_{min.} = \Delta N = -kN_o\Delta t \tag{A4.8}$$

Consider the rate of mineralization of soil organic matter (*SOM*) in a vineyard soil over 1 year. The value of N_o is large, but the decay coefficient is much smaller than that of a green manure. Suppose that for *SOM*, $k = 0.005$ yr^{-1} and the amount of soil organic N is 2000 t/ha. From equation A4.8, over 1 year, we have

$$N_{min.} = 0.005 \times 2000 \tag{A4.9}$$

$$= 10 \text{ kg N/ha} \tag{A4.10}$$

Values of k depend not only on the type of organic material, but also on changes in soil conditions such as occur during repeated cycles of air-drying and rewetting, or freezing and thawing.

Appendix 5 *Insoluble Hydroxides of Micronutrient Metals and How Complex Formation Increases Their Availability to Plants*

We use the reactions of hydrated Fe^{3+} as an example. The first step in the hydrolysis of this ion is

$$Fe(H_2O)_6^{3+} + H_2O \longleftrightarrow FeOH(H_2O)_5^{2+} + H_3O^+ \tag{A5.1}$$

The hydroxy-Fe ion continues to hydrolyze, releasing H^+ ions, until finally ferric hydroxide $Fe(OH)_3$ precipitates. The reaction is driven to the right by the presence of OH^- ions, which combine with the H_3O^+ ions. The overall process is summarized as

$$Fe^{3+}(\text{soln}) + 3OH^-(\text{soln}) \longleftrightarrow Fe(OH)_3 \text{ (solid)} \tag{A5.2}$$

Equation A5.2 is a dissolution-precipitation reaction for which the *solubility constant* K_{sp} is defined by

$$K_{sp} = \frac{(Fe^{3+})(OH^-)^3}{(Fe(OH)_3)} \tag{A5.3}$$

Taking the activity of pure, solid $Fe(OH)_3$ as 1, equation A5.3 becomes

$$K_{sp} = (Fe^{3+})(OH^-)^3 \tag{A5.4}$$

Two points follow from equation A5.4:

1. The smaller the value of K_{sp}, the lower is the solubility of the compound, and the lower the concentration of Fe in solution.
2. The higher the concentration of OH^- ions, the lower the concentration of Fe in solution.

The availability of metal cations can be increased by the formation of soluble complexes, using specific chelating agents (section 2.3.4.2). For example, Fe^{3+} forms very stable complexes with several organic ligands. If the soluble organic complex is more stable at a given pH than $Fe(OH)_3$, Fe will be maintained in solution and not precipitated. The stability constant of the complex determines which chelating agent is chosen to form a soluble complex with Fe, to increase its availability at high pH, especially on calcareous and chalky soils. Ethylenediamine tetra-acetic acid (EDTA) is a synthetic chelating agent that forms stable complexes with most micronutrient cations. Figure A5.1 shows that the complex $FeEDTA^-$ is normally more stable than $Fe(OH)_3$ to around pH 9, above which $Fe(OH)_3$ is more stable. But in calcareous soils, the crossover occurs around pH 8 because of the increasing stability of the $CaEDTA^{2-}$ complex at high pH. Thus, chelating agents that have a higher affinity for Fe than for Ca, such as EDDHA (ethylene diamine di(*o*-hydroxyphenylacetic acid)), should be used on calcareous soils. Natural chelating agents for Fe and other metals occur in the soil organic matter and leaf litter (box 1.5).

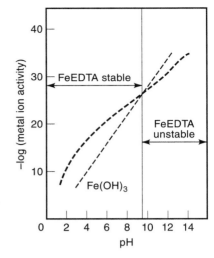

Figure A5.1 Stability of FeEDTA with pH change in soil (White 1997). Reproduced with permission of Blackwell Science Ltd.

Appendix 6   *Measuring a Soil's Redox Potential*

The redox potential E_h is measured as the electrical potential difference between a platinum (Pt) electrode and a reference electrode (usually a calomel electrode) placed in the soil. The expression for E_h (in millivolts) is

$$E_h = E_7^0 + \frac{59}{n} \log_{10} \frac{(Ox)}{(Red)} \tag{A6.1}$$

In this equation, n is the number of electrons transferred per mole of substance reduced, and (Ox) and (Red) are the activities of the oxidized and reduced forms of the substance, respectively. Since H^+ ion activity affects the redox equilibrium (see equations 5.12–5.15), the reference potential E_7^0 is defined at pH 7. Clearly, systems in an oxidized state have a high value of E_h, whereas reduced systems have a low value (sometimes even negative; see fig. 5.9).

The range of E_h for aerobic soils is 300–800 mV. Under anaerobic conditions, the critical E_h below which damage to vine roots occurs is 100mV. This might occur regularly in poorly drained subsoils in wet winters or transiently in duplex soils in winter because of a perched watertable developing at the top of the B horizon. The former condition can be alleviated by permanent subsoil drainage, the latter by deep ripping the soil at the time of vineyard establishment (chapter 7). Reproducible and meaningful E_h measurements can be made only on waterlogged soils in situ or on soil samples that have been taken in such a way that O_2 is excluded up to, and during, the time of measurement. Measurements on soil samples that have been dried, sieved, and shaken in water are worthless for identifying the redox status of the soil in its field state.

Appendix 7 *Free Energy Change and the Definition of Soil Water Potential*

Suppose a small quantity of water is added to a dry, nonsaline soil at a constant temperature. This water is distributed so that the force of attraction between it and the soil is as large as possible, and the free energy of the water is reduced to a minimum. More water added to the soil is held by progressively weaker forces. Finally, with the last drop of water necessary to saturate the soil, the free energy of the soil water approaches that of pure water (no solutes) at the same temperature and pressure. The reference water must be "free"—that is, not affected by any forces other than gravity. The energy status of soil water is then defined in terms of the difference in free energy between 1 mole of water in the soil and 1 mole of pure, free water at a standard temperature and pressure, and a fixed, reference height.

The free energy per mole of a substance defines its *chemical potential.* So for water in soil, we define the *soil water potential* ψ by the equation

$$\psi = \mu_w(\text{soil}) - \mu_w{}^\circ \ (\text{standard state}) = RT \ln e/e^\circ \tag{A7.1}$$

where μ_w = chemical potential (free energy/mol) of water; e/e° = the ratio of vapor pressure of water in the soil to vapor pressure of pure water at a standard temperature (298 K) and pressure (1 atmosphere); R = the Universal Gas Constant (8.314 Joule/Kelvin/mol); and T = the absolute temperature (K). The reference height used for measuring soil water potentials is usually the soil surface. Note that for water at the standard temperature and pressure, and reference height, the maximum value of ψ is obtained at $e/e^\circ = 1$ for which $\psi = 0$.

Appendix 8 *Measuring Saturated and Unsaturated Hydraulic Conductivity of Soil*

The *ring infiltrometer* is a metal cylinder 30–60 cm wide that is driven a short distance into the soil. Water is ponded in the ring and when steady state infiltration is achieved, the infiltration rate *IR* is measured as

$$IR = \frac{\text{Volume of water}}{\text{Area of ring} \times \text{Time}} \tag{A8.1}$$

Water movement into the soil, especially if the soil is initially dry, will be both lateral and vertical, under the influence of ψ_m and ψ_g gradients. The infiltrometer will therefore overestimate the vertical *IR* (gravity effect only), which equals K_s, unless an outer "buffer ring" filled with water is used, as in the *double-ring infiltrometer*. In this case, the infiltration rate is measured from the inner ring only.

The *disc permeameter* or tension infiltrometer operates on a similar principle to the ring infiltrometer, but it is more versatile because water can be applied at pressure heads $(-p)$ from 0 to -120 mm. Water infiltrates from a circular disc (radius r_o) covered by a porous membrane of air entry pressure >2 kPa. A thin layer of fine sand is spread on the soil surface under the disc to ensure good contact. The rate of water loss (Q) from the permeameter is measured at steady state infiltration. The value of K is calculated from an approximate formula that allows for the lateral suction effect on infiltration, that is,

$$K = \frac{Q/\pi r_0^2}{1 + 4/\pi r_0 \alpha} \tag{A8.2}$$

The value of α ranges from 0.0.05 to 0.5 cm^{-1} for clay loams to sands. The disc permeameter is used to obtain the relationship between K and $-p$ for the initial part of the curves shown in figure B6.7.1. K values measured close to saturation can vary 10- to 100-fold within an area of 1 ha because of soil variability. A pressure head of -20 mm (-0.2 kPa) is recommended for measuring K_s, because at this small head, the effect of very large pores is eliminated and the measured K_s is less variable spatially. Even so, at least 10 measurements should be made per hectare to obtain a mean K_s of acceptable precision.

Appendix 9

Table A9.1   ***Calculation of Stored Soil Water at* FC *(mm/m depth of soil) for a Soil Profile with Varying Texture with Depth***

Soil Texture	Depth at Which θ is Measured (m)	θ at *FC*[a] (m³/m³)	Effective Depth Interval, Δz (m)	Stored Water per Depth Interval, Δz (mm)
Sandy loam	0.2	0.30	0.3	90[b]
Sandy loam	0.4	0.25	0.2	50
Clay loam	0.6	0.42	0.2	84
Clay loam	0.8	0.42	0.2	84
Clay	1.0	0.42	0.1	42
			Total effective depth of measurement = 1 m	Total profile stored water $S = 350$ mm

[a]Measured by neutron probe or TDR (section 6.4.3)
[b]$\theta \times \Delta z \times 1000$

Appendix 10 *Volume of Irrigation Water Applied by Different Irrigation Methods*

The soil is a sandy clay loam for which the *DAW* is 0.05 m³/m³ or 50 mm/m depth. The desired *SWD* range is between -80 and -130 mm/m depth of soil. The data refer to water required per ha and per vine, for *RDI* by different irrigation methods, on vines planted at different densities.

(a) For *overhead sprinklers, flood,* or *furrow* methods, the volume is calculated per ha. It is assumed that the vine roots extend to the midrow and midvine po-

Table A10.1   *Irrigation Requirement for* **RDI** *by Overhead Sprinkler, Flood, or Furrow*

DAW (mm/m Depth)	Depth of Soil Drying (m)	Depth of Irrigation Water (mm)	Volume per m² of Soil Surface (L)	Volume of Water per ha (ML)[a]
50	0.4	20	20	0.2
	0.6	30	30	0.3
	0.8	40	40	0.4

[a]Megaliters

sitions. When the lower limit of *SWD* is reached, the amount of water needed to replenish the *DAW* for different rooting depths (hence different depths of soil drying) is as given in table A10.1.

(b) For irrigation by *microjets* or *microsprinklers*, the water is applied within the row, and the wetted soil area is restricted (ca. 1 m²). The volume of water required depends on the number of jets, which is determined by the planting density (vine by row spacing) and the rooting depth. The example is for *RDI* on a soil of *DAW* = 0.05 m³/m³ (50 mm/m depth). The irrigation requirements are listed in table A10.2.

Note that microjets and minisprinklers conserve more water per ha than overhead sprinklers, flood, and furrow.

(c) For *drip irrigation* (section 6.6.1), the spread of water under a dripper depends on how dry the soil is and how the soil's hydraulic conductivity K changes with θ. Initially, when the soil is dry, matric forces pull the water laterally as well as vertically. However, as the soil gets wet, gravity becomes the dominant force. For short wetting times and soils of relatively low K, the wetting zone under a dripper will be approximately hemispherical. For longer times, the zone becomes more elongated downward. In soils of high K value (sands and sandy loams), the wetting zone may become approximately cylindrical (with the long axis vertical).

Table A10.2   *Irrigation Requirement for* **RDI** *by Microjet or Minisprinkler*

Vine by Row Spacing (m × m)	No. of Vines per ha	Wetted Area per Vine (m²)	Volume of Water per Vine (L) for a Rooting Depth of			Volume of Water (ML) per ha for a Rooting Depth of		
			0.4 m	0.6 m	0.8 m	0.4 m	0.6 m	0.8 m
Dense 1 × 2	5,000	1	20	30	40	0.1	0.15	0.2
Less dense[a] 1.5 × 3	2,222	1	20	30	40	0.044	0.067	0.089

[a]If the number of jets were increased to 2 per vine, the quantities of water required per ha would be similar to that required in the denser planting.

Table A10.3   *Irrigation Requirement under Drip Irrigation*

Vine by row spacing (m × m)	No. of vines per ha	Diameter (m) of Wetted Soil Volume for a Wetting Depth of		Volume of Water per Vine (L) for a Wetting Depth of		Volume of Water (ML) per ha for a Wetting Depth of	
		0.4 m	0.6 m	0.4 m	0.6 m	0.4 m	0.6 m
$K_s = 0.02$ m/hr							
1 × 2	5,000	0.57	0.66	5.2	10.1	0.026	0.051
1.5 × 3	2,222	0.57	0.66	5.2	10.1	0.011	0.022
$K_s = 0.04$ m/hr							
1 × 2	5,000	0.46	0.52	3.3	6.4	0.017	0.032
1.5 × 3	2,222	0.46	0.52	3.3	6.4	0.007	0.014

The effects of these soil properties are quantified by a formula (Dasberg and Or, 1999) for calculating the diameter d of wet soil, for a given depth of wetting, as follows:

$$d = 1.32\left(\frac{zq}{K_s}\right)^{0.33}$$ (A 10.1)

where z is the depth of wetting required, q is the irrigation rate, and K_s is the saturated hydraulic conductivity. Note that the diameter of wet soil increases as K_s decreases. The d value is obtained by substituting appropriate values for z, q, and K_s in equation A10.1. The depth of wetting is determined by the depth of the main fibrous root system of the vine. In the following examples, the shape of wetted soil volume V_s is assumed to be cylindrical and is calculated as follows:

$$V_s = \pi\left(\frac{d}{2}\right)^2 z$$ (A10.2)

The volume of water required for *RDI* is then the product of V_s and *DAW*. For this soil example (*DAW* = 0.05 m³/m³), with two possible K_s values, the volume of water required has been calculated assuming one dripper per vine (4 L/hr) for two depths of wetting and two vine spacings. The results are given in table A10.3.

An increase in K_s causes the diameter of wetted soil to decrease for a given depth of wetting, which means that the volume required per vine for a particular *RDI* is decreased; also the quantities of water required would be doubled if there were two drippers per vine. Note that the water required per ha to achieve a desired *RDI* is less under drip irrigation than by microjets, and much less than under overhead sprinklers or flood irrigation.

Appendix 11   *Irrigation Schedule Using Weather Data and Crop Coefficients for a Vineyard in Southeastern Australia*

The aim is not to stress the vines from bud burst to fruit set (*SWD* kept within the *RAW* range), but to apply some stress from fruit set to veraison (*SWD* kept within the *DAW* range). The effective rooting depth is 0.6 m. The soil is a clay loam with *RAW* = 80 mm/m depth and *DAW* = 50 mm/m depth (48 and 30 mm, respectively, for the effective rooting depth).

Stage of Growth	Week No.	Rain, P (mm/week)	E_p (mm/week)	Crop Coeff., C_c	E_a (mm/week)	$P - E_a$ (mm/week)	SWD (mm/0.6 m)	Drainage (−) or Irrigation (+) (mm)
Bud burst	1	10	20	0.2	4	6	0	−6
(1 Oct.)	2	0	22	0.2	4.4	−4.4	−4.4	0
	3	0	25	0.2	5	−5	−9.4	0
	4	2	30	0.2	6	−4	−13.4	0
	5	0	32	0.2	6.4	−6.4	−19.8	0
	6	4	30	0.4	12	−8	−27.8	0
	7	10	35	0.4	14	−4	−31.8	0
	8	0	36	0.4	14.4	−14.4	−46.2	0
Irrigate to return *SWD* to 0; volume required = 46 mm, or 46 L/m², adjusted according to application method (see appendix 10)							0	+46
Flowering	9	4	38	0.4	15.2	−11.2	−11.2	0
	10	12	38	0.4	15.2	−3.2	−14.4	0
	11	5	40	0.4	16	−11	−25.4	0
Fruit set	12	0	42	0.3	12.6	−12.6	−38	0
	13	0	46	0.3	13.8	−13.8	−51.8	0
	14	0	49	0.3	14.7	−14.7	−66.5	0
	15	12	40	0.3	12	0	−66.5	0
Veraison (February)	16	0	49	0.3	14.7	−14.7	−81.2	0
Irrigate to return *SWD* to −48 mm; volume required = 33 mm or 33 L/m² (adjusted as done previously)							−48.2	+33

Appendix 12 *The Relationship Between* EC$_{1:5}$ *and* EC$_e$ *for a Range of Soil Textural Classes*

To convert $EC_{1:5}$ values to EC_e values, multiply the $EC_{1:5}$ value by the factor in table A12.1. The factor decreases as texture increases because, per unit weight of soil, a clay holds more water at saturation than a sand: the dilution effect on the soil solution is therefore relatively smaller in the clay soil.

Table A12.1   *Conversion Factors for* EC$_{1:5}$ *Values According to Soil Texture*

Soil Texture Class	$EC_e = EC_{1:5} \times$ Factor
Loamy sand	13
Loam	11
Sandy clay loam	8.5
Light clay	6.5
Heavy clay	5

Appendix 13   *Hot Water Treatment of Rootlings and Cuttings*

Hot water treatment is recommended to reduce the numbers of nematodes and phylloxera on the roots of dormant rootlings or to control infections of crown gall and phytoplasmas (associated with grapevine yellows) in dormant cuttings (Caudwell et al. 1997, Nicol et al. 1999)). The cuttings must be fully dormant before treatment. The treatment involves

Rootstocks and *V. vinifera* cuttings	Hot water at 50°C for 30 minutes
Rootlings (bare roots)	Hot water at 50°C for 5 minutes

The bundles of cuttings and rootlings should be hydrated and brought to ambient temperature before being immersed in containers of hot water. Immediately after treatment, the bundles are plunged into clean water at ambient temperature for 20–30 minutes. Those that are to be kept in cold storage before being used may be dipped in the fungicide Chinosol (8-hydroxyquinoline sulfate) for control of pathogens.

Appendix 14   *The Heat Summation Method of Classifying Climates for Viticulture*

This climate classification, also known as the Heat Degree Days (HDD) system, is based on the knowledge that the grapevine neither grows actively nor matures its fruit at mean daily temperatures $<10°C$ (50°F). The steps in the calculation of HDD for a particular location follow:

1. Calculate the mean daily temperature for each month during the growing season. The mean daily temperature is given by

$$\text{Mean daily temperature} = \frac{\text{Daily maximum} - \text{Daily minimum temperature}}{2}$$

(A14.1)

These mean temperatures should be based on at least 10 years of measurement.

2. Subtract the base temperature of 10°C and multiply by the number of days in each month. Sum the products for the number of months of the growing season (usually 7 months). The result is expressed as degree-days, or heat-summation units. To convert from temperatures in Celsius to Fahrenheit (used in the United States), multiply by 1.8.

In California, where this system was developed, five regions based on heat summation were recognized, as shown in table A14.1 (Amerine and Winkler 1944).

This classification system has been modified by others, such as Gladstones (1992), who developed a system in which monthly mean temperatures $>19°C$ were ignored in the calculation. Another system was developed by Smart and Dry (1980), who found that the mean January temperature was well correlated with HDD for places of similar continentality. The continentality index of climate

Table A14.1   *Regions and Temperature Ranges of the HDD Climate Classification*

Region	HDD Units (°Celsius)	HDD Units (°Fahrenheit)	Qualitative Description
I	944–1390	1700–2500	Very cool
II	1391–1670	2501–3000	Cool
III	1671–1940	3001–3500	Moderately cool
IV	1941–2220	3501–4000	Warm
V	>2220	>4001	Very warm to hot

(CTL) is measured by the difference between the mean January and mean July temperatures (for the southern hemisphere). The HDD system only roughly accounts for the climatic factors that affect vine growth and grape ripening, but it has been widely used to guide the expansion of grape production into cooler areas in the New World.

Appendix 15   *Conversion Factors for SI Units (International System of Units) and Non-SI Units, Including American Units and SI Abbreviations[a] Used in This Book*

To Convert Column 1 into Column 2, Multiply by	Column 1, SI Unit	Column 2, non-SI Unit	To Convert Column 2 into Column 1, Multiply by
Length, area, and volume			
3.28	meter, m	foot, ft	0.304
39.4	meter, m	inch, in	0.0254
3.94×10^{-2}	millimeter, mm	inch, in	25.4
2.47	hectare, ha	acre, ac	0.405
0.265	liter, L	gallon	3.78
9.73×10^{-3}	cubic meter, m^3	acre-inch, acre-in	102.8
8.11×10^{-4}	cubic meter, m^3	acre-foot, acre-ft	1.233×10^3
35.3	cubic meter, m^3	cubic foot, ft^3	2.83×10^{-2}
0.811	megaliter, ML	acre-foot, acre-ft	1.233
Mass			
2.20×10^{-3}	gram, g	pound, lb	454
2.205	kilogram, kg	pound, lb	0.454
1.102	tonne, t	ton (U.S), ton	0.907
Quantities per unit area			
0.893	kilogram/hectare, kg/ha	pound/acre, lb/ac	1.12
0.446	tonnes/hectare, t/ha	tons (U.S.)/ac, ton/ac	2.24
0.107	liter/hectare, L/ha	gallon/ac	9.35

To Convert Column 1 into Column 2, Multiply by	Column 1, SI Unit	Column 2, non-SI Unit	To Convert Column 2 into Column 1, Multiply by
Miscellaneous			
(9/5°C) + 32	Celsius, °C[b]	Fahrenheit, °F	5/9 (°F − 32)
9.90	megapascal, MPa	atmosphere	0.101
10	siemen/meter, S/m	millimho/centimeter, mmho/cm	0.1

[a]mega (M) = $\times 10^6$; kilo (k) = $\times 10^3$; deci (d) = $\times 10^{-1}$; centi (c) = $\times 10^{-2}$; milli (m) = $\times 10^{-3}$; micro (μ) = $\times 10^{-6}$; nano (n) = $\times 10^{-9}$

[b]To convert from °Celsius to Kelvin, add 273.

References

Amerine, M. A., and A. J. Winkler. 1944. Composition and quality of musts and wines of California grapes. *Hilgardia* 15: 493–575.

Baize, D. 1993. Total and "active" calcium carbonate. In *Soil science analyses. A guide to current use*, trans. Graham Cross, pp. 61–68. New York: Wiley.

Barbeau, G., R. Morlat, C. Asselin, A. Jacquet, and C. Pinard. 1998. Comportement du cépage Cabernet Franc dans différentes terroirs du val de Loire. Incidence de la précocité sur la composition de la vendange en année climatique normale (exemple de 1998). *Journal International de Sciences de Vigne et du Vin* 32: 69–81.

Boyer, J. D., and T. K. Wolf. 2000. Development and preliminary validation of a Geographic Information System approach to vineyard site suitability assessment in Virginia. Proceedings 5th International Symposium on Cool Climate Viticulture and Oenology, Workshop 16 "Site Selection and Vineyard Planning," pp. 1–10. Adelaide: Australian Society of Viticulture and Oenology.

Buckerfield, J. C., and K. A. Webster. 1996. Earthworms, mulching, soil moisture and grape yields. *Australian and New Zealand Wine Industry Journal* 11: 47–53.

Buckerfield, J. C, and K. A. Webster. 2001. Managing earthworms in vineyards—improve incorporation of lime and gypsum. *The Australian Grapegrower and Winemaker,* Technical issue no. 449a: 55–61.

Busby, J. 1825. *A treatise on the culture of the vine and the art of making wine.* Australia: R. Howe, Government Printer.

Campostrini, F., E. A. C. Costanini, F. Maltori, and G. Nicolini. 1996. Effect of "Terroir" on quanti-qualitative parameters of "vino nobile di Montepulciano." In *Les Terroirs Viticoles. Concept, Produit, Valorisation*, pp. 461–468. Montpellier: Institut National de la Recherche Agronomique.

Cass, A. 1998. Measuring and managing chemical impediments to growth. *Australian Grapegrower and Winemaker*, July: 13–16.

Cass, A. 1999. Interpretation of some physical indicators for assessing soil physical fertility. In *Soil analysis: An interpretation manual*, eds. K. I. Peverill, L. A. Sparrow, and D. J. Reuter, pp. 95–102. Melbourne: CSIRO Publications.

Cass, A., B. Cockcroft, and J. M. Tisdall. 1993. New approaches to vineyard and orchard soil preparation and management. In *Vineyard development and redevelopment*, ed. P. F. Hayes, pp. 18–24. Adelaide: Australian Society of Viticulture and Oenology.

Cass, A., D. Maschmedt, and J. Chapman. 1998. Managing physical impediments to root growth. *The Australian Grapegrower and Winemaker*, June: 13–17.

Caudwell, A., J. Larrue, E. Boudon-Padieu, and G. D. McLean. 1997. Flavescence dorée elimination from dormant wood of grapevines by hot-water treatment. *Australian Journal of Grape and Wine Research* 3: 21–25.

Christensen, L. P., A. N. Kasimatis, and F. L. Jensen. 1978. *Grapevine nutrition and fertilization in the San Joaquin Valley.* Oakland, Calif.: Division of Agriculture and Natural Resources, University of California.

Coombe, B. G., and P. R. Dry, eds. 1988a. *Viticulture. Volume 1 Resources.* Adelaide: Winetitles.

Coombe, B. G., and P. R. Dry, eds. 1988b. *Viticulture. Volume 2 Practices.* Adelaide: Winetitles.

Conradie, W. J. 1980. Seasonal uptake of nutrients by Chenin blanc in sand culture. I Nitrogen. *South African Journal of Enology and Viticulture* 1: 59–65.

Darlington, D. 1999. Burgundy of the west. *Via* September/October: 36–41.

Dasberg, S., and D. Or. 1999. *Drip irrigation.* Berlin: Springer-Verlag.

Dean, R., and F. Berwick. 2001. We live in a marketing world—except in Europe. *Australian and New Zealand Wine Industry Journal* 16 (6): 65–67.

Dry, P. R., and B. R. Loveys. 1998. Factors influencing grapevine vigour and the potential for control with partial root zone drying. *Australian Journal of Grape and Wine Research* 4: 140–148.

Elliot-Fisk, D. 1993. Viticultural soils of California, with special reference to the Napa Valley. *Journal of Wine Research* 4: 67–77.

Emerson, W. W. 1991. Structural decline in soils, assessment and prevention. *Australian Journal of Soil Research* 29: 905–921.

Enjalbert, H. 1983. *Les grands vins de St Emilion, Pomerol et Fronsac.* Paris: Editions Bardi.

FAO. 1998. *The world reference base for soil resources (WRB).* World Soil Resources Report No. 84. Rome: Food and Agriculture Organization of the United Nations.

FAO-Unesco. 1988. *Soil map of the world. Revised legend.* World Resources Report 60. Rome: Food and Agriculture Organization of the United Nations.

Flaherty, D. L., L. P. Christensen, W. T. Lanini, J. J. Marois, P. A. Phillips, and L. T. Wilson. 1992. *Grape pest management,* 2nd ed., Oakland, Calif.: Division of Agriculture and Natural Resources, University of California.

Freeman, B. M., and R. E. Smart. 1976. Research note: A root observation laboratory for studies with grapevines. *American Journal of Enology and Viticulture* 27: 36–39.

Gishen, M., B. Dambergs, M. Esler, L. Francis, P. Iland, R. Johnstone, and A. Kambouris. 2001. Objective measures of grape and wine quality. Proceedings of the 11th Australian Wine Industry Technical Conference, p. 32, Adelaide: Australian Society of Viticulture and Oenology.

Gladstones, J. S. 1992. *Viticulture and the environment. A study of the effects of environment on grapegrowing and wine qualilty with emphasis on present and future areas for growing winegrapes in Australia.* Adelaide: Winetitles.

Goldspink, B. H., ed. 1996. *Fertilisers for wine grapes.* Perth: Agriculture Western Australia.

Goodwin, I. 1995. *Irrigation of vineyards.* Tatura, Victoria: Institute of Sustainable Irrigated Agriculture.

Greenland, D. J., P. J. Gregory, and P. H. Nye. 1998. Land resources and constraints to crop production. In *Feeding a world population of more than eight billion people. A challenge to science,* eds. J. C. Waterlow, D. G. Armstrong, L. Fowden, and R. Riley, pp. 39–55. New York: Oxford University Press.

Hall, D. G. M., M. J. Reeve, A. J. Thomasson, and V. F. Wright. 1977. *Water retention, porosity and density of field soils.* Technical Monograph No. 9, Harpenden, England: Soil Survey of England and Wales.

Halliday, J. 1993. Climate and soil in Australia. *Journal of Wine Research* 4: 19–34.

Hancock, J. 1999. Feature review. *Terroir. The role of geology, climate, and culture in the making of French wines* by J. E. Wilson, 1998. *Journal of Wine Research* 10:43–49.

Hudson, N. 1995. *Soil Conservation*, 3rd ed. London: Batsford.

Iacono, F., D. Porro, F. Campostrini, and A. Bersan. 2000. Site evaluation and selection to optimise quality of wine. Proceedings 5th International Symposium on Cool Climate Viticulture and Oenology, pp. 1–5, Session 1A. Adelaide: Australian Society of Viticulture and Oenology.

Iland, P., and P. Gago. 1997. *Australian wine from the vine to the glass.* Adelaide: P. Iland Wine Publishers.

Isbell, R. F. 1996. *The Australian soil classification.* Australian Soil and Land Survey Handbook. Melbourne: CSIRO Publishing.

Itami, R. M., J. Whiting, K. Hirst, and G. Maclaren. 2000. Use of analytical hierarchy process in cool climate GIS site selection for wine grapes. Proceedings 5th International Symposium on Cool Climate Viticulture and Oenology, Section 1B Climate and Crop Estimation, pp. 1–8. Adelaide: Australian Society of Viticulture and Oenology.

Jackson, R. E. 2000. *Wine science: Principles, practice, perception*, 2nd ed. London: Academic Press.

Jenny, H. 1941. *Factors of soil formation.* New York: McGraw-Hill.

Johnson, H. 1994. *The world atlas of wine*, 4th ed. New York: Simon and Schuster.

Jones, J. B., Jr., ed. 1999. *Soil analysis handbook of reference methods.* Soil and Plant Analysis Council Inc., Boca Raton, Fla.: CRC Press.

Jury, W. A., W. R. Gardner, and W. H. Gardner. 1991. *Soil physics*, 5th ed. New York: Wiley and Sons.

Kennedy, A. 2001. An Australian case study. Proceedings 11th Australian Wine Industry Technical Conference, p. 33. Adelaide: Australian Society for Viticulture and Oenology.

Kliewer, W. M. 1991. Methods for determining the N status of vineyards. In *Nitrogen in grapes and wine*, ed. J. M. Rantz, pp. 133–147. Davis, Calif.: American Society for Enology and Viticulture.

Klute, A., ed. 1986. *Methods of soil analysis. Part 1. Physical and mineralogical methods*, 2nd ed., monograph no. 9. Madison, Wisc.: American Society of Agronomy/Soil Science Society of America.

Lambert, G., and J. Kashiwagi. 1978. *Soil survey of Napa County, California.* Washington, D.C.: United States Department of Agriculture, Soil Conservation Service.

Linacre, E. 1992. *Climate data and resources. A reference and guide.* London: Routledge.

Livingston, J. 1998. The geology of fine wine. *California Wild.* California Academy of Sciences Quarterly.

Lohnertz, O. 1991. Soil nitrogen and the uptake of nitrogen in grape vines. In *Nitrogen in grapes and wine*, ed. J. M. Rantz, pp. 1–11, Davis, Calif.: American Society for Enology and Viticulture.

Martin, D. 2000. The search for terroir—a question of management. Proceedings 5th International Symposium on Cool Climate Viticulture and Oenology, Session 1A, pp. 1–4. Adelaide: Australian Society of Viticulture and Oenology.

Maschmedt, D., R. Fitzpatrick, and A. Cass. 2002. Key for identifying categories of vineyard soils in Australia. CSIRO Land and Water Technical Report 30/02, Adelaide, Australia: CSIRO.

Mason, B. 1966. *Principles of geochemistry*. New York: Wiley and Sons.

May, P. 1994. *Using grapevine rootstocks. The Australian perspective*. Adelaide: Winetitles.

McDonald, R. D., R. F. Isbell, J. G. Speight, J. Walker, and M. S. Hopkins. 1990. *Australian soil and land survey field handbook*, 2nd ed. Melbourne: Inkata Press.

McKenzie, D. C., ed. 1998. *Soilpak for cotton growers*, 3rd ed. Orange: New South Wales Agriculture.

McKenzie, D. C. 2000. Soil survey—an important preliminary to vineyard design. Proceedings 5th International Symposium on Cool Climate Viticulture and Oenology, pp. 1–4. Melbourne; Australian Society for Viticulture and Oenology.

Mitchell, P. D., and I. Goodwin. 1996. *Micro-irrigation of vines and fruit trees*. Melbourne: Agriculture Victoria.

Mullins, M. G., A. Bouquet, and L. E. Williams. 1992. *Biology of the grapevine*. Cambridge: Cambridge University Press.

Myburgh, P., A. Cass, and P. Clingeleffer. 1998. Root systems and soils in Australian vineyards—an assessment. Adelaide: Cooperative Research Centre for Soil and Land Management.

Neja, R. A., R. S. Ayers, and A. N. Kasimatis. 1978. Salinity appraisal of soil and water for successful production of grapes, Leaflet 21056. Davis, Calif.: Division of Agricultural Science, University of California.

Nicol, J. M., G. R. Stirling, B. J. Rose, P. May, and R. van Heeswijck. 1999. Impact of nematodes on grapevine growth and productivity: Current knowledge and future directions, with special reference to Australian viticulture. *Australian Journal of Grape and Wine Research* 5: 109–127.

Noble, A. C., R. A. Arnold, J. Buechsenstein, E. J. Leach, J. O. Schmid, and P. M. Stern. 1987. Modification of a standardized system of wine aroma technology. *American Journal of Enology and Viticulture* 38: 143–146.

Northcote, K. H., J. S. Russell, and C. B. Wells. 1954–59. *Soils and land use in the Barossa district, South Australia*. Adelaide: CSIRO Division of Soils.

Penn, C. 2001. What is quality? An American perspective. *Australian and New Zealand Wine Industry Journal* 16 (3): 58–59.

Pinchon, M. 1996. Allocution. In *Les terroirs viticoles. Concept, produit, valorisation*. Montpellier: Institut National de la Recherche Agronomique.

Pitts, D., M. Bianchi, and C. Clark. 1995. Scheduling microirrigations for winegrapes using CIMIS. Proceedings 5th International Microirrigation Congress, pp. 792–798. Orlando, Fla.: American Society of Agricultural Engineers Publication 4.

Pomerol, C. 1989. *The wines and winelands of France. Geological journeys*. London: Robertson McCarta.

Pongracz, D. P. 1983. *Rootstocks for grapevines*. Cape Town: David Philip Publications.

Possingham, J. V., and J. Groot Obbink. 1971. Endotrophic mycorrhiza and the nutrition of grape vines. *Vitis* 10: 120–130.

Priestley, C. H. B., and R. J. Taylor. 1972. On the assessment of surface heat flux and evaporation using large-scale parameters. *Monthly Weather Review* 100: 81–92.

Pudney, S., T. Proffitt, A. Brown, and D. Willoughby, D. 2001. Soil moisture sensor demonstration—Barossa Valley 2000/2001. *Australian Grapegrower and Winemaker Annual Technical Issue:* 76–84.

Rankine, B. C., J. C. M. Farnachon, E. W. Boehm, and K. M. Celler. 1971. Influence of grape variety, climate and soil on grape composition and quality of table wines. *Vitis* 10: 33–50.

Raupach, M., and B. M. Tucker. 1959. The field determination of soil reaction. *Journal of the Australian Institute of Agricultural Science* 25: 129–133.

Rayment, G. E., and F. R. Higginson. 1992. *Australian laboratory handbook of soil and water chemical methods.* Australian Soil and Land Survey Handbook. Melbourne: Inkata Press.

Rhoades, J. D., and S. Miyamoto. 1990. Testing soils for salinity and sodicity. In *Soil testing and plant analysis*, 3rd ed., R. L. Westerman, pp. 299–336. Madison, Wisc.: Soil Science Society American Book Series no. 3.

Richards, D. 1983. The grape root system. *Horticultural Reviews* 5: 127–168.

Robinson, J., ed. 1999. *The Oxford companion to wine*, 2nd ed. Oxford: Oxford University Press.

Robinson, J. B., M. T. Treeby, and R. A. Stephenson. 1997. Fruits, vines and nuts. In *Plant analysis. An interpretation manual*, 2nd ed., D. J. Reuter and J. B. Robinson, pp. 349–382. Melbourne: CSIRO Publishing.

Schaller, K. 1991. Groundwater pollution by nitrate in viticultural areas. In *Nitrogen in grapes and wine*, ed. J. M. Rantz, pp. 12–22. Davis, Calif.: American Society for Enology and Viticulture.

Seguin, G. 1986. "Terroirs" and pedology of wine growing. *Experientia* 42: 861–873.

Seguin, G. 1972. Répartition dans l'espace du système rediculaire de la vigne. *Compte Rendus Academe Science Paris*, 274 D: 2178–2180.

Smart, R. E. 2001. Where to plant and what to plant. *Australian and New Zealand Wine Industry Journal* 16 (4): 48–50.

Smart, R. E., and P. R. Dry. 1980. A climatic classification for Australian viticultural regions. *Australian Grapegrower and Winemaker* 196: 8–16.

Smart, R., and M. Robinson. 1991. *Sunlight into wine. A handbook for winegrape canopy management.* Adelaide: Winetitles.

Smith, M. 1992. *Expert consultation on revision of FAO methodologies for crop water requirements.* Rome: Food and Agriculture Organization of the United Nations.

Soil Survey Division Staff. 1993. *Soil survey manual*, 3rd ed. Washington, D.C.: United States Department of Agriculture, National Soil Survey Center. USDA-NRCS Soil Survey Division Data National STATSGO Database.

Soil Survey Staff. 1996. *Keys to soil taxonomy*, 7th ed. Washington, D.C.: United States Department of Agriculture.

Sposito, G. 1989. *The chemistry of soils.* New York: Oxford University Press.

Stace, H. C. T., G. D. Hubble, R. Brewer, K. H. Northcote, J. R. Sleeman, M. J. Mulcahy, and E. G. Hallsworth. 1968. *A handbook of Australian soils.* Adelaide, South Australia: Rellim.

Tesic, D., D. J. Woolley, E. W. Hewett, and D. J. Martin. 2002. Environmental effects on cv Cabernet Sauvignon (*Vitis vinifera* L.) grown in Hawke's Bay, New Zealand. II Development of a site index. *Australian Journal of Grape and Wine Research* 8: 27–35.

Tinker, P. B., and P. H. Nye. 2000. *Solute movement in the rhizosphere.* Oxford: Oxford University Press.

van Leeuwen, C. 1989. *Carte des sols du vignoble de Saint-Emilion.* St Emilion: Syndicat Viticole de Saint Emilion.

van Leeuwen, C. 1996. La notion de terroir viticole dans le Bordelais. Occasional paper, École Nationale d'Ingenieurs des Travaux Agricoles. Bordeaux: Faculté d'Oenologie, Bordeaux University.

van Leeuwen, C., and I. Merouge. 1998. Les sols viticoles de St. Emilion et de Pomerol. In *Guidebook for A3 Tour across the Southwest of France from Bordeaux to Carcassonne* (coordinators D. Arrouays and C. Mathieu), 16th World Congress of Soil Science. Montpellier: Institut National de la Recherche Agronomique.

van Leeuwen, C., R. Renard, O. Leriche, C. Molot, and J.-P. Soyer. 1998. Le fonctionnement de trois sols viticoles du Bordelais: Conséquences sur la croissance de la vigne et sur le potential oenologique du raisin en 1997. *Revue Francaise d'Oenologie* 170: 28–32.

Verdegaal, P. S. 1999. Preplanting decisions in establishing a vineyard. San Joaquin County: University of California Farm Advisor.

Wahl, K. 2000. The relationship between soil and the sensorial characteristics of wines. Proceedings 5th International Symposium on Cool Climate Viticulture and Oenology, Session 5A. Adelaide: Australian Society of Viticulture and Oenology.

Walker, A. 1999. Ampelography short course—rootstocks. Davis, Calif.: University of California.

Weber, E. 1992. Spacing and trellis decisions. Viticulture Notes. Napa County, Calif.: University of California Cooperative Extension.

Webster, R., and M. A. Oliver. 2000. *Geostatistics for environmental scientists*. London: Wiley and Sons.

Wetherby, K. 2000. *Soil description book*. Clare, South Australia: K. G. and C. V. Wetherby.

White, R. E. 1997. *Principles and practice of soil science. The soil as a natural resource*, 3rd ed. Oxford: Blackwell Science.

Williams, B. 2000. The fate of herbicides used in viticulture. Part 2. *Australian Viticulture* 4 (1): 11–17.

Williams, L. E. 1996. Effects of soil water content and environmental conditions on vine water status and gas exchange of *Vitis vinifera L.* cv Chardonnay. Proceedings 1st Colloque International "Les Terroirs Viticoles," pp. 161–163. Angers, France: Institut National de la Recherche Agronomique.

Williams, L. E. 1999. Using crop coefficients to schedule irrigations in the San Joaquin Valley—practical applications. Kearney Agricultural Center Grape Day Notes. Fresno, Calif.: University of California.

Williams, L. E., and P. J. Biscay. 1991. Partitioning of dry weight, nitrogen, and potassium in Cabernet Sauvignon grapevines from anthesis until harvest. *American Journal of Enology and Viticulture* 42: 113–117.

Wilson, J. E. 1998. *Terroir. The role of geology, climate, and culture in the making of French wines*. London: Mitchell Beazley.

Winkler, A. J., J. A. Cook, W. M. Kliewer, and L. A. Lider. 1974. *General viticulture*. Berkeley, Calif.: University of California Press.

Wolpert, J., A. Walker, E. Weber, L. Bettiga, R. Smith, and P. Verdegaal. 1994. Rootstocks and phylloxera: A status report for coastal and northern California. Viticulture Notes no. 6. Napa County: University of California, Cooperative Extension Service.

Index

accessory minerals, 36, 40–41. *See also* aluminium oxides and iron oxides

acid rain, 121, 126

adsorption, 27, 51, 96, 143. *See also* sorption
  nonspecific and specific, 95

aggregates. *See* soil aggregates

albedo, 73

Alsace-Lorraine, 137

aluminum
  exchangeable, 99, 177
  hydrolysis of, 101. *See also* pH buffering capacity
  oxides, 41

allophane 28, 37

alluvium, 13
  soils on, 224, 231, 233, 239, 248

American Viticultural Areas, 4, 227, 241

ammonium, 79, 82, 87–88, 97–98, 115
  fertilizers, 121. *See also* nitrogen
  volatilization of, 123

anion exchange capacity, 41, 96

anthocyanin pigments, 230, 233, 235
  and grape color, 228

Appellation d'Origine Contrôllée system, 4, 224, 227

available water capacity, 149, 162, 174, 229. *See also* plant available water
  deficit available water, 162, 212, 248
  readily available water, 162, 212, 232, 248

bacteria, 44, 89, 129. *See also* decomposers
  facultative anaerobes, 135

obligate anaerobes, 136

Barossa Valley, South Australia, 47, 224
  soils of, 245
  and wine quality, 227–228

base saturation, 101

Beaujolais region, 12, 134, 239

Bellarine Peninsula, Victoria, 215

biological N_2 fixation
  amounts of N fixed, 85
  biochemistry of, 86
  factors affecting, 86
  free-living organisms, 85
  symbiotic fixation, 84–85

biomass. *See also* nutrient cycling
  microbial, 42
    measurement of, 48, 252
  dynamics of change, 52
  soil, 43

biotite, 34

Black Earth, 243

blue-green algae or Cyanobacteria, 45, 85, 204

Bordeaux
  Mixture, 132, 138, 194
  region, 55, 195, 230
    soils of, 232–235

Bordelais. *See* Bordeaux region

boron, 131

Bourgogne. *See* Burgundy region

Brown Calcareous Soil. *See* Calcareous Brown Soil

Brown Forest Soil, 7